S0-DJW-517

MASTER TUBE SUBSTITUTION HANDBOOK

TAB EDITORIAL STAFF

TAB BOOKS
Blue Ridge Summit, Pa. 17214

FIRST EDITION

FIRST PRINTING— AUGUST 1976

Copyright c 1976 by TAB BOOKS

Printed in the United States
of America

Hardbound Edition: International Standard Book No. 0-8306-6870-8

Paperbound Edition: International Standard Book No. 0-8306-5870-X

Library of Congress Card Number: 76-24782

Preface

This handbook is intended for use by TV technicians, experimenters, hobbyists, do-it-yourselfers, or anyone else who might have their hand in electronics. Everyone who has trouble keeping up with the vast number of tubes in production today will find this handbook an invaluable aid in those inevitable "tight spots" when the correct tube type is not readily available. A TV technician will soon discover that this handbook will equal a super tube caddy filled with countless replacements that could not otherwise be carried on each home call.

For the sake of convenience, this handbook is broken down into three sections. Section 1 contains vacuum tube substitutions. Section 2 is comprised of tube base diagrams, and Section 3 covers both monochrome and color picture tube substitutions.

An effort has been made to adhere to basic tube types throughout this guide. Suffix letters such as G, GT, A, B, etc. are included. In some cases, a suffix may be an indication of a difference in tube size. Space considerations aside, you may assume that these tubes are interchangeable in nearly all applications. A brief summary of the most common tube suffixes follows:

—G signifies a glass bulb and an octal base.
—GT refers to a T-9 straight-sided glass bulb and an octal base.
—A, B, C, D, E, F, etc., assigned in that order signifies a later and modified version which can be substituted for any previous

version. but preferably not vice versa. The suffix assigned does not convey any information regarding the modification.
—X means the tube base is composed of a special low-loss material.
—Y specifies a base made of a special intermediate-loss material.

Additionally. S and SF suffixes may be ignored when considering substitutions. Also. tubes with the suffix P (indicating an extra grid). may be substituted for one with the suffix T. but not vice versa. The first modification of a tube. shown by the suffix A. can always be assumed to be interchangeable with the original.

Much of the information in this handbook has been taken from its predecessor (TAB book number 570). with additional material from RCA and EIMAC. We would like to thank RCA and EIMAC for supplying us with the update information necessary to complete this work. Our special thanks go to RCA Advertising/Sales Promotion Manager. Mr. A. D. Ricketti and Bill Orr of EIMAC for their cooperation in assiting and supplying information to TAB Books.

Mike Fair
Service Editor
TAB Editorial Staff

CONTENTS

SECTION 1
VACUUM-TUBE
SUBSTITUTIONS

This section covers vacuum tubes of all types, excluding picture tubes which are located in *Section 3*. Receiving tubes, transmitting tubes, industrial tubes, commercial tubes, and foreign tubes are listed in order on the following pages.

To use this information begin by referring to the first column to locate the number of the tube in question. The system used to list tubes in the first column is as follows: *Number-letter-number* designations come first (e.g. 0A2). This first group of tubes is the standard way of numbering tubes and is by far the largest portion of tubes in this section. Next comes *numbered* tubes (e.g. 7025). The final portion of this section covers tubes with *letter-number* designations such as CC88.

We have made the list as complete as possible, with all tubes listed and all variations thereof. That is, we have listed as many known versions of each tube with suffixes such as A, B, GT, GTX, and so on. This information may aid you in several ways. For one, it may tell you that the tube under consideration is superseded by a newer version. Another way that it might help you is to help you determine which version fits the circuit best. For example, if you have at your disposal a L6G and a 6L6GC, use the G-version in an old piece of equipment and save the GC-version for later equipment.

The third column contains the *preferred* substitute while the fourth column contains the *good* substitute. This does not mean that the good substitute will not function properly. It does mean that when you have both at your disposal you should use the preferred substitute.

Original	Base Diagram	Preferred Substitution	Good Substitution
0A2	5BO	6073, 6626, 0A2WA	150C1
0A2WA	5BO	0A2, 6073, 6626	
0A3	4AJ	0A3A, VR75	
0A3A	4AJ	0A3, VR75	
0A4G	4V	RL1267	
0A5	6CB		
0B2	5BO	0B2WA, 6074	VR90
0B2WA	5BO	0B2, 6074	VR90
0B3	4AJ	VR90	
0C2	5BO		0A3
0C3	4AJ	0C3A, VR105	
0C3A	4AJ	0C3, VR105	
0D3	4AJ	0D3A	VR150
0D3A	4AJ	0D3	VR150
0Y4	4BU		
0Z4	4R	0Z4A, 0Z4G	
0Z4A	4R	0Z4G, 0Z4	
0Z4G	4R	0Z4A, 0Z4	
1A3	5AP		
1A4	4M	1A4P, 1B4	1C4, 1K4
1A4P	4M		
1A4T	4K		
1A5	6X	1T5, 1A5GT	1C5, 1Q5
1A5GT	6X	1A5, 1T5	1C5, 1Q5
1A6	6L	1C6	
1A7	7Z	1A7GT	1B7, 1C7, 1D7, 1B7GT
1A7GT	7Z	1A7	1B7, 1C7, 1D7, 1B7GT
1AB5	5BF		
1AC5	8CP	1V5	1AD5, 1W5
1AD2	12GV	1AD2A	1AJ2
1AD2A	12GV	1AD2	1AJ2
1AD4	1AD4		1AH4, 1AK40

Original	Base Diagram	Preferred Substitution	Good Substitution
1AD5	8CP	1W5	1AC5
1AE4	6AR		1L4, 1U4
1AE5	9U		
1AF4	6AR	1U4	1AM4
1AF5	6AU		1AR5, 1S5
1AG4	512AX		6526
1AG5	1AG5		1AJ5, 1AK5
1AH4	1AD4	1AK4	5875
1AH5	6AU	1AF5	1AR5
1AJ2	12EL		1AD2
1AJ4	6AR	1AF4	1AM4, 1T4
1AJ5	1AG5		1AK5
1AK4	1AD4	1AH4	5875
1AK5	1AG5		1AJ5
1AM4	6AR	1T4	1AF4
1AQ5	7AT		1R5
1AR5	6AU	1R5	1AF5
1AS5	6BW	1U5	1DN5
1AU3	3C		1J3, 1K3, 1B3 1N2, 1G3
1AX2	9Y	1BK2	1S2
1AY2	1AY2	1AY2A	
1AY2A	1AY2	1AY2	
1B3	3C	1N2, 1B3GT	1AU3, 1G3, 1J3, 1K3
1B3GT	3C	1B3, 1N2	1AU3, 1G3, 1K3, 1J3
1B4	4M	1A4, 1B4P	34
1B4P	4M	1B4, 1A4	34
1B5	6M	1B5/25S	
1B5 25S	6M	1B5	
1B7	7Z	1B7GT, 1A7	1C7, 1D7
1B7GT	7Z	1B7, 1A7	1C7, 1D7
1B8	8AW		1D8

Original	Base Diagram	Preferred Substitution	Good Substitution
1BC2	9RG	1BC2A	1BH2, 1BH2A, 1AX2, 1BK2
1BC2	9RG	1BC2	1BH2A, 1BH2, 1AX2, 1BK2
1BH2	9RG	1BH2A	1BC2, 1BC2A, 1AX2, 1BK2
1BK2	9Y	1AX2	1S2
1BL2	1AY2		1AY2
1BY2	12HZ		
1C3	5CF		
1C4	4K	1K4	1A4, 1B4
1C5	6X	1C5GT	
1C5GT	6X	1C5	
1C6	6L	1A6	
1C7	7Z	1D7, 1C7G, 1D7G	
1C7G	7Z	1C7, 1D7, 1D7G	
1C8	8CN		1E8
1C21	4V	359	
1D3	8DN		
1D5	5Y	1D5GP, 1D5GT	1E5
1D5GP	5Y	1D5, 1D5GT	1E5
1D5GT	5Y	1D5, 1D5GP	1E5
1D7	7Z	1C7, 1C7G, 1D7G	
1S7G	7Z	1D7, 1C7, 1C7G	
1D8	8AW	1D8GT	1B8
1D8GT	8AW	1D8	1B8
1DG3	8ND	1DG3A	
1DG3A	8ND	1DG2	
1DN5	6BW	1U5	1AS5
1DY4	7DK		
1E4	5S	1G4, 1G4GT	1H4
1E5	5Y	1D5, 1D5GP, 1D5GT, 1E5GP	
1E5GP	5Y	1E5, 1D5, 1D5GP, 1D5GT	

Original	Base Diagram	Preferred Substitution	Good Substitution
1E7GT	8C		
1E8	8CN		1C8
1F2	6AR		1L4
1F4	5K		33
1F5	6X	1F5G	1G5, 1J5, 1G5G, 1J5G
1F5G	6X	1F5	1G5, 1J5, 1G5G, 1J5G
1F6	6W		
1F7	7AF		
1G3	3C	1G3GT, 1G3GTA, 1B3, 1B3GT	1N2, 1AU3, 1K3, 1J3
1G3GT	3C	1G3, 1G3GTA, 1B3, 1B3GT	1N2, 1AU3, 1K3, 1J3
1G3GTA	3C	1G3, 1G3GT, 1B3, 1B3GT	1N2, 1AU3, 1K3, 1J3
1G4	5S	1G4GT, 1E4	1H4
1G5	6X	1J5, 1J5G, 1G5G	1F5, 1F5G
1G5G	6X	1G5, 1J5, 1J5G	1F5, 1F5G
1G6	7AB	1G6GT	1J6, 1J6G, 1J6GT
1G6GT	7AB	1G6	1J6, 1J6G, 1J6GT
1H2	9LX		1S2
1H4	5S	1H4G	1E4, 1G4, 1G4GT
1H5	5Z	1H5GT	
1H6	7AA	1H6G	
1H6G	7AA	1H6	
1J3	3C	1K3, 1K3A, 1N2, 1AU3	1B3, 1G3, 1B3GT, 1G3GTA
1J5	6X	1G5, 1G5G, 1J5G	1F5, 1F5G
1J5G	6X		
1J6	7AB	1J6G, 1J6GT	1G6, 1G6GT

Original	Base Diagram	Preferred Substitution	Good Substitution
1J6G	7AB	1J6, 1J6GT	
1J6GT	7AB	1J6, 1J6G	
1K3	3C	1K3A, 1AU3, 1N2, 1J3	1B3, 1B3GT, 1G3, 1G3GTA
1K3A	3C	1K3, 1AU3, 1N2, 1J3	
1K4	4K	1C4	1A4, 1B4, 1B4P
1K5	5Y	1M5	1D5, 1E5, 1D5GP, 1D5GT
1L4	6AR		1T4
1L6	7DC		1U6
1LA4	5AD	1LB4	
1LA6	7AK		1LC6
1LB4	5AD		1LA4
1LC5	7AO	1LN5, 1LG5	
1LC6	7AK		1LA6
1LD5	6AX		
1LE3	4AA	1LF3	
1LF3	4AA	1LE3	
1LG5	7AO	1LN5	1LC5
1LH4	5AG		
1LN5	7AO	1LG5	1LC5
1M3	8GV		1N3
1N2	3C		1B3, 1B3GT, 1G3, 1G3GT, 1G3GTA, 1AU3, 1K3, 1K3A, 1J3,
1N3	8GV		1M3
1N5	5Y	1P5, 1N5GT, 1P5GT	1D5, 1E5GT, 1D5GT, 1D5GP
1N5GT	5Y	1N5, 1P5, 1P5GT	1D5, 1D5GP, 1D5GT, 1E5, 1E5GP
1N6G	7AM		

Original	Base Diagram	Preferred Substitution	Good Substitution
1P5	5Y	1N5, 1P5GT, 1N5GT	1E5, 1E5GP, 1D5, 1D5GP, 1D5GT
1P5GT	5Y	1P5, 1N5, 1N5GT	1E5, 1E5GP, 1D5, 1D5GP, 1D5GT
1Q5	6AF	1C5, 1Q5GT, 1C5GT	1T5, 1T5GT
1Q5GT	6AF	1Q5, 1C5, 1C5GT	1T5, 1T5GT
1Q6	9RT		
1R4	4AH		
1R5	7AT		1AF5, 1AR5
1S2	9DT	1BK2, 1S2A/DY87	1AX2, 1H2
1S2A	9DT	1S2, 1BK2, 1S2A/DY87	1AX2, 1H2
1S4	7AV		
1S5	6AU		
1S6	8DA	1T6	
1T4	6AR	1L4	1U4
1T5	6X	1F5, 1T5GT, 1F5G	1A5, 1G5, 1J5, 1G5G, 1J5G
1T5GT	6X	1T5, 1F5, 1F5G	1A5, 1G5, 1J5, 1G5G, 1J5G
1T6	8DA	1S6	
1U4	6AR	5910	1T4, 1AF4
1U5	6BW	1DN5	1AS5
1U6	7DC		1L6
1V	4G	6Z3	
1V2	9U		
1V5	4G	1AC5	1W5, 1AD5
1W5	4G	1AD5	1AC5, 1V5
1X2	9Y	1X2A, 1X2B, 1X2C	1BK2, 1AX2
1X2A	9Y	1X2, 1X2B, 1X2C	1BK2, 1AX2
1X2B	9Y	1X2, 1X2A, 1X2C	1BK2, 1AX2
1X2C	9Y	1X2, 1X2B, 1X2A	1BK2, 1AX2

Original	Base Diagram	Preferred Substitution	Good Substitution
2A3	4D	5930	2DX4
2A5	6B		
2A6	6G		
2A7	7C		
2AF4	7DK	2AF4A, 2DZ4, 2AF4B	2T4
2AF4A	7DK	2AF4, 2AF4B, 2DZ4	2T4
2AF4B	7DK	2AF4, 2AF4A, 2DZ4	2T4
2AH2	12DG	2BU2	
2AS2	12EW	2AS2A	
2AS2A	12EW	2AS2	
2AV2	9U		2BA2
2AZ2	9Y	2BJ2	
2B7	7D		
2BA2	9MP		2AV2
2BJ2	9RT	2BJ2A	
2BJ2A	9RT	2BJ2	
2BN4	7EG	2BN4A	
2BN4A	7EG	2BN4	
2BU2	12JB	2AH2	
2C21	7BH	6J5, 6J5GT	1642
2C22	4AM	5670, 5670WA	7193
2C39A			
2C39WA			
2C50	8BD	6BL7, 6BL7GT, 6BL7GTA, 6BX7, 6DN7	
2C51	8CJ		5670
2C52	8BD	6EM7, 6EA7, 6GL7, 6SL7, 6SU7, 6SL7A, 6SL7GT, 6SL7GTY, 6SL7Y	2C50

Original	Base Diagram	Preferred Substitution	Good Substitution
2CN3	8MU	1CN3A	
2CN3A	8MU	2CN3	
2CW4	12AQ	2DS4	2EG4
2CY5	7EW	2EA5	2EV5
2D21	7BN	2D21W, 5727, 6012	
2D21W	7BN	2D21, 5727, 6012	
2DS4	12AQ	2CW4	2EG4
2DV4	12EA		
2DX4	7DK	2T4	2AF4, 2AF4A, 2AF4B
2DY4	7DK		2DX4
2DZ4	7DK	2AF4, 2AF4A, 2AF4B	2T4
2E5	6R		
2EA5	7EW	2EV5	
2EG4	12AQ		2CW4, 2DS4
2EN5	7FL		
2ER5	7FP	2ES5, 2GK5, 2FQ5, 2FQ5A	
2ES5	7FP	2ER5, 2GK5, 2FQ5, 2FQ5A	
2EV5	7EW	2CY5	2EA5
2FH5	7FP	2ES5	2ER5, 2FQ5, 2FY5, 2GK5
2FQ5	7FP	2FQ5A, 2FY5, 2GK5	2ER5, 2ES5
2FQ5A	7FP	2FQ5, 2FY5, 2GK5	2ER5, 2ES5
2FS5	7GA	2GU5	
2FV6	7FP	2CY5	
2FY5	7FP	2FQ5, 2FQ5A, 2GK5	2ER5, 2ES5

Original	Base Diagram	Preferred Substitution	Good Substitution
2GK5	7FP	2FQ5, 2FQ5A, 2FY5	2ER5, 2ES5
2GU5	7GA	2FS5	
2HA5	7GM	2HM5, 2HK5, 2HQ5	
2HK5	7GM	2HM5, 2HA5, 2HQ5	
2HM5	7DK	2HA5, 2HQ5, 2HK5	
2T4	7DK	2AF4, 2AF4A, 2DZ4	
2T24		3C24	
2V2	8FV		3C2
2X2	4AB	2Y2	
2X1000A			
2X3000F			
2Y2	4AB	2X2	
2-01C			
2-25A			
2-50A			
2-150D			
2-2000A			
3A2	9DT	3A2A	
3A2A	9DT	3A2	
3A3	8EZ	3B2, 3A3A, 3A3B, 3A3C	3AW3, 3CN3, 3CN3A, 3CN3B, 3CU3, 3CU3A, 3CZ3, 3CZ3A, 3DB3, 3CY3, 3DC3, 3DJ3
3A3A	8EZ	3B2, 3A3, 3A3B, 3A3C	3AW3, 3CN3, 3CN3A, 3CN3B,

Original	Base Diagram	Preferred Substitution	Good Substitution
			3CU3.
			3CU3A.
			3CZ3.
			3CZ3A.
			3DB3. 3CY3.
			3DC3. 3DJ3
3A3B	8EZ	3A3. 3A3A. 3A3C. 3B2	3AW3. 3CN3. 3CN3A. 3CN3B. 3CU3. 3CZ3. 3CZ3A. 3DB3. 3CY3. 3DC3. 3DJ3
3A3C	8EZ	3A3. 3A3A. 3A3B. 3B2	3AW3. 3CN3. 3CN3A. 3CN3B. 3CU3. 3CU3A. 3CZ3. 3CZ3A 3DB3. CY3. 3DC3. 3DJ3
3A4	7BB		3Q4. 3W4
3A5	7BC		
3A8	8AS	3A8GT	
3A8GT	8AS	3A8	
3AF4	7DK	3AF4A. 3DZ4	3DX4
3AF4A	7DK	3DZ4. 3AF4	3DX4
3AL5	6BT		
3AT2	12FV	3AT2A. 3AT2B	3AW2. 3AW2A. 3BN2. 3BN2A. 3BW3. 3BS3A. 3BT2
3AT2A	12FV	3AT2. 3AT2B	3AW3. 3AW2A.

(continued)

Original	Base Diagram	Preferred Substitution	Good Substitution
			3BN2, 3BN2A, 3BW2, 3BS2A, 3BT2
3AT2B	12FV	3AT2, 3AT3A	3AW2, 3AW2A, 3BN2, 3BN2A, 3BN2, 3BN2A, 3BW2, 3BS2A, 3BT2
3AU6	7BK	3BA6, 3DK6	3BC5, 3BZ6, 3CB6, 3CE5, 3CF6
3AW2	12EW	3AT2, 3AT2A, 3AT2B, 3AW2A	3BN2, 3BN2A, 3BW2, 3BS2A, 3BT2
3AW2A	12EW	3AW2, 3AT2, 3AT2A, 3AT2B	3BN2, 3BN2A, 3BW2, 3BS2, 3BT2
3AW3	8EZ	3A3, 3A3A, 3A3B, 3A3C, 3DJ3	3B2, 3CN3, 3CN3A, 3CN3B, 3CU3, 3CU3A, 3CZ3, 3CZ3A, 3DB3, 3CY3, 3DC3, 3DJ3
3B2	8GH	3A3, 3A3A, 3A3B, 3A3C, 3DJ3	3AW3, 3CN3, 3CN3A, 3CN3B,

Original	Base Diagram	Preferred Substitution	Good Substitution
			3CU3, 3CU3A, 3CZ3, 3CZ3A, 3DB3, 3CY3, 3DC3, 3DJ3
3B4WA	7CY		
3B5	7AQ		3C5, 3Q5, 3Q5GT
3BA6	7BK	3AU6	3BC5, 3BZ6, 3CB6, 3CF6, 3DK6
3BC5	7BD	3BZ6, 3CB6, 3CF6, 3DK6, 3BC5, 3CE5	3AU6, 3BA6
3BE6	7CH	3BY6, 3CS6	
3BL2	12HK	3BL2A	3BM2, 3AT2, 3AT2A, 3AT2B, 3AW2, 3AW2A
3BL2A	12HK	3BL2	3BM2, 3AT2, 3AT2A, 3AT2B, 3AW2, 3AW2A
3BM2	12HK		3AW2, 3AW2A, 3BN2, 3BN2A, 3AT2, 3AT2A, 3AT2B
3BN2	12FV	3BN2A	3AW2, 3AW2A, 3BM2, 3AT2, 3AT2A, 3AT2B
3BN2A	12FV	3BN2	3AW2, 3AW2A,

(continued)

Original	Base Diagram	Preferred Substitution	Good Substitution
			3BM2, 3AT2, 3AT2A, 3AT2B
3BN4	7EG	3BN4A	
3BN4A	7EG	3BN4	
3BN6	7DF		
3B52	12HY	3B52A, 3B52B, 3BT2, 3BW2	3AT2, 3AT2A, 3BT2A, 3AT2B, 3AW2, 3AW2A
3BS2A	12HY	3BW2, 3BT2, 3BS2, 3BS2B, 3BT2A	3AT2, 3AT2A, 3AT2B, 3AW2, 3AW2A
3BS2B	12HY	3BS2, 3BS2A, 3BT2, 3BT2A, 3BW2	3AT2, 3AT2A, 3AT2B, 3AW2, 3AW2A
3BT2	12HY	3BW2, 3BS2, 3BS2A, 3BS2B, 3BT2A	3AT2, 3AT2A, 3AT2B, 3AW2, 3AW2A
3BT2A	12HY	3BT2, 3BS2, 3BS2A, 3BS2B, 3BW2	3AT2, 3AT2A, 3AT2B, 3AW2, 3AW2A
3BU8	9FG	3BU8A, 3HS8, 3KF8, 3GS8	
3BU8A	9FG	3BU8, 3HS8, 3KF8, 3GS8	
3BW2	12HY	3BT2, 3BS2, 3BS8A	3AT2, 3AT2A, 3AT2B, 3AW2, 3AW2A
3BX6	9AQ	3BY7	
3BY6	7CH	3CS6, 3BE6	

Original	Base Diagram	Preferred Substitution	Good Substitution
3BY7	9AQ	3BX6	
3BZ6	7CM	3CB6, 3CF6, 3DK6	3AU6, 3BA6, 3BC5, 3DT6
3C4	6BX	3E5	3V4
3C5	7AQ	3C5GT, 3Q5, 3Q5GT	3B5
3C5GT	7AQ	3C5, 3Q5, 3Q5GT	3B5
3C24			
3C200			
3CA3	8MH	3CA3A	3A3, 3A3A, 3A3B, 3A3C, 3CU3 3CU3A, 3CZ3, 3CZ3A, 3CN3, 3CN3A, 3CN3B
3CA3A	8MH	3CA3	3A3, 3A3A, 3A3B, 3A3C, 3CU3, 3CU3A, 3CZ3, 3CZ3A, 3CN3, 3CN3A, 3CN3B
3CB6	7CM	3BZ6, 3CF6, 3DK6, 3AU6, 3BA6, 3BC5	
3CE5	7BD	3BC5	3AU6, 3BA6, 3BZ6, 3CB6, 3DK6
3CF6	7CM	3BZ6, 3CB6	3AU6, 3BA6, 3DK6, 3BC5
3CN3	8MU	3CN3A, 3CN3B	3A3, 3A3A,

(continued)

Original	Base Diagram	Preferred Substitution	Good Substitution
			3A3B, 3A3C, 3CU3, 3CU3A, 3CZ3, 3CZ3A
3CN3A	8MU	3CN3, 3CN3B	3A3, 3A3A, 3A3B, 3A3C, 3CU3, 3CU3A, 3CZ3, 3CZ3A
3CN3B	8MU	3CN3, 3CN3A	3A3, 3A3A, 3A3B, 3A3C, 3CU3, 3CU3A, 3CZ3, 3CZ3A
3CPN10A5		7815	
3CPX100A5		7815R	
3CPX1500A7			
3CS6	7CH		3BE6, 3BY6
3CU3	8MK	3CU3A	3A3, 3A3A, 3A3B, 3A3C, 3CN3, 3CN3A, 3CN3B, 3CZ3, 3CZ3A
3CU3A	8MK	3CU3	3A3, 3A3A, 3A3B, 3A3C, 3CN3, 3CN3A, 3CN3B, 3CZ3, 3CZ3A
3CV3	8EZ	3CV3A, 3DB3, 3CY3, 3DJ3	3AW3, 3A3, 3A3A, 3A3B, 3A3C,

24

Original	Base Diagram	Preferred Substitution	Good Substitution
			3CU3, 3CU3A
3CV3A	8EZ	3CV3, 3DB3, 3CY3, 3DJ3	3AW3, 3A3, 3A3A, 3A3B, 3A3C, 3CU3, 3CU3A
3CV1500A7			
3CV30000A1			
3CV30000A3			
3CV30000H3			
3CB50000A7			
3CW5000A1	8240		
3CW5000A3	8242		
3CW5000F1	8241		
3CW5000F3	8243		
3CW5000H3			
3CW10000A3			
3CW10000H3			
3CW20000A1			
3CW20000A3			
3CW20000A7			
3CW20000H3			
3CW20000H7			
3CW30000H3			
3CW30000H7			
3CW40000H3			
3CX3	8MT	3DA3, 3DH3	
3CX100A5	7289		
3CX100F5	8250		
3CX400U7	8961		
3CX1000A7	8283		
3CX1500A7	8877		
3CX2500A3	18161		

Original	Base Diagram	Preferred Substitution	Good Substitution
3CX2500F3		8251	
3CX2500H3			
3CX3000A1		8238	
3CX3000A7			
3CX3000F1		8239	
3CX3000F7		8162	
3CX5000A3			
3CX5000H3			
3CX10000A1		8158	
3CX10000A3		8159	
3CX10000A7		8160	
3CX10000H3			
3CX15000A3			
3CX15000A7			
3CX15000H3			
3CX20000A3			
3CX20000A7			
3CX20000H3			
3CY3	8MX	3DB3, 3CY3, 3DJ3	3A3, 3A3A, 3A3B, 3A3C, 3CU3, 3CU3A
3CY5	7EW	3EA5	3EV5
3CZ3	8EZ	3CZ3A	3AW3, 3A3, 3A3A, 3A3B, 3A3C, 3CU3, 3CU3A
3CZ3A	8EZ	3CZ3	3AW3, 3A3, 3A3A, 3A3B, 3A3C, 3CU3, 3CU3A
3D6	6BA		3LE4, 3LF4
3DA3	8MY	3DH3	

Original	Base Diagram	Preferred Substitution	Good Substitution
3DB3	8MX	3CY3	3A3, 3A3A, 3A3B, 3A3C, 3CU3, 3CU3A
3DC3	8MZ		
3DE6	7CM	3BZ6	
3DF3	8MT	3DF3A	
3DF3A	8MT	3DF3	
3DG4	5DE		
3DH3	8MY	3DA3	
3DJ3	8MX	3DB3, 3CY3	3A3, 3A3A, 3A3B, 3A3C, 3CU3, 3CU3A
3DK6	7CM	3BZ6, 3CB6, 3CF6	3AU6, 3BA6, 3BC5
3DR3	8NL	3DA3, 3DH3, 3DS3	
3DS3	8NL	3DA3, 3DH3, 3DR3	
3DT6	7EN	3DT6A	
3DT6A	7EN	3DT6	
3DX4	7DK		3AF4, 3DZ4, 3AF4A
3DY4	7DK	3DX4	3AF4, 3AF4A, 3DZ4
3DZ4	7DK	3AF4, 3AF4A	3DX4, 3DY4
3E5	6BX	3C4	3V4
3EA5	7EW	3EV5, 3CY5	
3EH7	9AQ	3EH7/XF183	3EJ7, 3JC6, 3JC6A, 3JD6, 3KT6
3EJ7	9AQ	3EJ7/XF184	3EH7, 3JC6, 3JC6A, 3JD6, 3KT6
3ER5	7FP	3ES5	
3ES5	7FP	3ER5	
3EV5	7EW	3EA5, 3CY5	

Original	Base Diagram	Preferred Substitution	Good Substitution
3F65		8185, 4-65A	
3FH5	7FP		3ER5, 3ES5, 3FQ5, 3FQ5A, 3GK5, 3FY5
3FQ5	7FP	3FQ5A, 3GK5, 3FY5	3ER5, 3ES5
3FQ5A	7FP	3FQ5, 3GK5, 3FY5	3ER5, 3ES5
3FS5	7GA	3GU5	
3FW7	8LM		3FX7
3FX7	8LK		3FW7
3FY5	7FP	3GK5, 3FQ5, 3FQ5A	3ER5, 3ES5
3GK5	7FP	3FQ5, 3FQ5A, 3FY5	3ER5, 3ES5
3GS8	8LW	3HS8, 3BU8	3KF8
3GU5	7GA	3FS5	
3H		7289, 151J	
3HA5	7GM	3HM5, 3HK5, 3HQ5	
3HC		3CX100A5, 151JYY	
3HK5	7GM	3HA5, 3HM5, 3HQ5	
3HM5	7GM	3HA5, 3HQ5, 3HK5	
3HM6	9PM	3HT6, 3JC6, 3JC6A	3EJ7, 3EH7
3HQ5	7GM	3HA5, 3HM5, 3HK5	
3HS8	9FG	3GS8, 3BU8	3KF8
3HT6	9PM	3HM6, 3JC6, 3JC6A	3EJ7, 3EH7
3JC6	9PM	3JC6A, 3JD6	3HM6, 3HT6
3JC6A	9PM	3JC6, 3JD6	3HM6, 3HT6
3JD6	9PM	3JC6, 3JC6A	3HM6, 3HT6

Original	Base Diagram	Preferred Substitution	Good Substitution
3KF8	9FG	3BU8. 3HS8. 3GS8	
3KT6	9PM	3JD6	3JC6. 3JC6A
3LE4	6BA	3LF4	3D6
3LF4	6BA	3LE4	3D6
3M-R24	7CM	3DK6	
3M-V7	7CM	3BZ6	
3Q4	7BA	3W4	3A4. 3Z4
3Q5	7AQ	3Q5G. 3Q5GT. 3C5. 3C5GT	3B5
3Q5G	7AQ	3Q5. 3Q5GT. 3C5. 3C5GT	3B5
3Q5GT	7AQ	3Q5. 3C5. 3Q5G. 3C5GT	3B5
3S4	7BA	3Q4. 3W4	3Z4
3S035T		5867A	
3V4	6BX		3C4. 3E5
3W4	7BA	3Q4	3S4. 3Z4
3Z4	7BA	3W4	3Q4. 3S4
3-200A3		592	
3-400Z		8163	
3-500Z			
3-1000Z		8164	
4AU6	7BK	4BA6	4BC5. 3BZ6. 4CB6. 4CE5. 3CF6
4AV6	7BT		
4BA6	7BK	4AU6	4BC5. 4BZ6. 4CB6. 4DK6. 4CF6. 4DE6
4BC5	7BD	4BZ6. 4CB6. 4CF6. 4DK6	4AU6. 4BA6
4BC8	9AJ	4KN8. 4BZ8	4BQ7. 4BK7. 4BQ7A. 4BZ7

Original	Base Diagram	Preferred Substitution	Good Substitution
4BE6	7CH	4CS6	
4BL8	9AE	4BL8/XCF80	5GH8, 5EA8, 5CQ8, 5U8
4BN6	7DF		
4BQ7	9AJ	4BQ7A, 4BZ7, 4BC8, 3BZ8	4KN8
4BQ7A	9AJ	4BQ7, 4BZ7, 4BC8, 4BZ8	4KN8
4BS8	9AJ	4BC8, 4BQ7, 4BQ7A, 4BX8, 4BZ7	4BX8, 4KN8
4BU8	9FG	4BU8A, 4HS8, 4KF8	4GS8
4BU8A	9FG	4BU8, 4HS8, 4KF8	4GS8
4BX8	9AJ	4BS8	4BC8, 4BQ7, 4BQ7A, 4BZ7, 4BZ8, 4KN8
4BY6	7CH	4CS6	
4BZ6	7CM	4CB6, 3CF6, 4DE6, 4DK6	4AU6, 3BA6, 4BC5
4BZ7	9AJ	4BC8, 4BQ7, 4BQ7A	4BZ8, 4KN8
4BZ8	9AJ	4BC8, 4BQ7, 4BQ7A, 4BZ7	4KN8
4CB6	7CM	4BZ6, 4CF6, 4DE6, 4DK6	4AU6, 4BA6, 4BC5
4CE5	7BD	4BC5	4AU6, 4BA6, 4DK6, 4BZ6, 4CB6, 4DE6
4CF6	7CM	4BZ6, 4CB6, 4DE6, 4DK6	4AU6, 4BA6, 4BC5
4CN15A			
4CPX250K		8590	

Original	Base Diagram	Preferred Substitution	Good Substitution
4CS6	7CH		4BE6
4CS250R			
4CV1500B			
4CV8000A			
4CV20000A			
4CV35000A			
4CV50000E			
4CV50000J			
4CV100000C		8351	
4CV250000A			
4CW800B			
4CV800F			
4CW2000A		8244	
4CW10000A		8661	
4CW25000A			
4CW50000E			
4CW50000J			
4CW100000D			
4CW100000E			
4CW250000A			
4CX125C			
4CX125F			
4CX250B		7203	
4CX250BC		8957	
4CX250FG		8621	
4CX250K		8245	
4CX250M		8246	
4CX250R		7580W	
4CX300A		8167	
4CX300Y		8561	
4CX350A		8321	
4CX350F		8322	
4CX300B		8904	
4CX600F			

Original	Base Diagram	Preferred Substitution	Good Substitution
4CX600J		8809	
4CX600JA		8921	
4CX1000A		8168	
4CX1000K		8352	
4CX1500A			
4CX1500B		8660	
4CX3000A		8169	
4CX5000A		8170	
4CX5000J		8909	
4CX5000R		8170W	
4CX10000D		8171	
4CX10000J		8171	
4CX15000A		8281	
4CX15000J		8910	
4CX35000C		8349	
4CY5	7EW	4-125A	
4D21			
4D21A		4-125A	
4DE6	7CM	4BZ6, 4CB6, 4CF6	4AU6, 4BA6, 4BC5, 4CE5
4DK6	7CM	4BZ6, 4CF6, 4CB6	4AU6, 4BA6, 4BC5
4DT6	7EN	4DT6A	
4DT6A	7EN	4DT6	
4E27A		5-125B	
4EH7	9AQ	4EH7/LF183	4EJ7, 4JC6, 4JC6A, 4KT6, 4JD6, 4HM6, 4HT6
4EJ7	9AQ	4EJ7/LF184	4EH7, 4KT6, 4HT6, 4HM6, 4JD6, 4JC6, 4JC6A

Original	Base Diagram	Preferred Substitution	Good Substitution
4ES8	9DE	4KN8, 4ES8/XCC189	4BC8, 4BQ7, 4BQ7A, 4BZ7, 4BZ8
4EW6	7CM	4GM6, 4LU6	4BZ6, 4DK6, 4CB6, 4DE6
4F15R		7034, 4X150A	
4F17R		8172, 4X150G	
4F20R		7609	
4F21		4-125A	
4F84		7843	
4FQ5	7FP	4FQ5An 4GK5, 4FY5	
4FQ5A	7FP	4FQ5, 4GK5, 4FY5	
4FS7	9MP	4HG8	
4FY5	7FP	4GK5, 4FQ5, 4FQ5A, 4FY5	
4GJ7	9QA	4GJ7/XCF801	4GX7
4GK5	7FP	4FQ5, 4FQ5A	4FY5
4GM6	7CM	3EW6, 4LU6	4BZ6, 4CB6, 4DE6, 4DK6
4GS7	9GF	4LJ8	5FG7
4GS8	9LW	4BU8, 4HS8	4KF8
4GX7	9AQ	4GJ7	
4GZ5	7CV		
4H1355M		7034, 4X150A	
4H1600M		7203, 4CX25B	
4HA5	7GM	4HA5/PC900, 4HQ5, 4HK5, 4HM5	
4HA7	12FQ	4HC7	
4HC		7203, 4CX250B, 1G0M	
4HC7	12FR	4HA7	

Original	Base Diagram	Preferred Substitution	Good Substitution
4HG8	9MP	4FS7	
4HK5	7GM	4HA5, 4HM5, 4HQ5	
4HM5	7GM	4HA5, 4HQ5, 4HK5	
4HM6	9PM	4HT6	4EH7, 4EJ7, 4JC6, 4JC6A, 4JD6, 4KT6
4HQ5	7GM		4HA5, 4HM5, 4HK5
4HS8	9FG	4GS8, 4BU8	4KF8
4HT6	9PM	4HM6	4EH7, 4EJ7, 4JC6, 4JC6A, 4JD6, 4KT6
4JC6	9PM	4JC6A, 4JD6, 4KT6	4EJ7, 4EH7
4JC6A	9PM	4JC6, 4JD6, 4KT6	4EJ7, 4EH7
4JD6	9PM	4JC6, 4JC6A, 4KT6	4EJ7, 4EH7
4JH6	7CM	4BZ6, 4CB6, 4DK6, 4DE6	
4JK6	7CM	4JL6, 4LU6	
4JL6	7CM	4JK6, 4LU6	
4K84		2-450A	
4KE8	9DC		
4KF8	9FG	4BU8, 4GS8, 4HS8	
4KN8	9AJ	4KN8/4RHH8, 4BZ8	4BC8, 4BQ7, 4BQ7A, 4BZ7
4KT6	9PM	4JD6, 4JC6, 4JC6A	4EH7, 4EJ7
4LJ8	9GF	4GS7	
4LU6	7CM	4EW6, 4GM6	
4MK8	9FG		
4PR60B		8252	

Original	Base Diagram	Preferred Substitution	Good Substitution
4PR60C		8252W	
4PR65A		8187	
4PR125A		8247	
4PR250C		8248	
4PR400A		8188	
4PR1000A		8189	
4PR1000B			
4RHH2	9AJ	4BQ7, 4BQ7A, 4BZ7	
4RHH8	9AJ	4KN8	
4S016T		4-123A	
4S040T		5D22, 4-250A	
4T10R		7289	
4T16		100TL	
4T17		100TH	
4T25R		8172, 4X150G	
4W300B		8249	
4W20000A		8173	
4X150A		7034	
4X150G		8172	
4X500A			
4-400A		7527	
4-400B		7527	
5AM8	9CY		
5AN8	9DA		
5AQ5	7BZ		
5AR4	5DA	5AR4/GZ34	
5AS4	5T	5AS4A, 5U4G, 5U4GB	5AU4, 5V3, 5V3A
5AS4A	5T	5AS4, 5U4G, 5U4GB	5AU4, 5V3, 5V3A
5AS9	9DS		
5AT8	9DW		5BE8, 5CL8
5AU4	5T	5V3, 5V3A	5AW4, 5U4G, 5U4GB

Original	Base Diagram	Preferred Substitution	Good Substitution
5AV8	9DZ	5B8	
5AW4	5T	5AS4. 5AS4A. 5U4G. 5U4GB	5V3. 5V3A
5AX4	5T	5R4. 5R4GB. 5R3GY. 5R4GYB	5AZ4. 5U4G. 5U4GB
5AZ4	5T		5AX4. 5U4G. 5U4GB
5B8	9EC	5AV8	
5BC3	9QJ	5BC3A	
5BC3A	9QJ	5BC3	
5BE8	9EG	5BR8	5CL8
5BK7	9AJ	5BK7A	4BC8. 4BQ7. 4BQ7A. 4BZ7
5BK7A	9AJ	5BK7. 5BQ7. 5BQ7A	4BC8. 4BQ7. 4BQ7A. 4BZ7
5BQ7	9AJ	5BQ7A. 5BK7. 5BZ7	
5BQ7A	9AJ	5BQ7. 5BK7. 5BZ7	
5BR8	9FA	5FV8. 5BE8	5CL8
5BS8	9AJ	5BQ7. 5BQ7A. 5BZ7	
5BT8	9FE		
5BW8	9HK		
5CG4	5L	5V4. 5V4A. 5AU4. 5Z4	
5CG8	9GF	5AT8	5BE8. 5BR8. 5CL8. 5FG7. 5FV8
5CL8	9FX	5CL8A. 5FV8	5BE8. 5BR8. 5FG7
5CL8A	9FX	5CL8. 5FV8	5BE8. 5BR8. 5FG7
5CM6	9CK	5CZ5	

Original	Base Diagram	Preferred Substitution	Good Substitution
5CM8	9FZ		5CR8. 5KZ8
5CQ8	9GE		5GH8. 5EA8. 5U8. 5GH8A
5CR8	9GJ	5KZ8	5CM8
5CX1500A			
5CX300A			
5CZ5	9HN		5CM6
5D22		4-250A	
5DB4	5T	5AS4. 5AS4A	5U4G. 5U4GB
5DH8	9EG		5BE8. 5BR8. 5CL8. 5FV8. 5BM8
5DJ4	8KS		5DN4
5EA8	9AE	5GH8. 5GH8A. 5U8	5CQ8
5ES8	9AJ	5ES8/YCC189	
5EU8	9JF		
5EW6	7CM	5GM6	
5F15R		7034. 4X150A	
5F16R		7609	
5F17R		8172. 4X150G	
5F20RA		7203. 4CX250B	
5F22		5D22. 4-250A	
5F22A		6156	
5F23		8438. 4-440A	
5F23A		7527. 4-440B	
5F25R		4CX250FG	
5F35R		8321. 4CX350A	
5FG7	9GF	5FV8. 5BR8	5DH8
5FV8	9FA	5FG7. 5BR8	5BE8. 5CL8
5GH8	9AE	5GH8A. 5EA8. 5U8	5CQ8
5GH8A	9AE	5GH8. 5EA8. 5U8	5CQ8
5GJ7	9QA	5GX7. 5GJ7 LCF801	

Original	Base Diagram	Preferred Substitution	Good Substitution
5GM6	7CM	5EW6	
5GS7	9GF	5LJ8	
5GX6	7EN	5HZ6	
5GX7	9QA	5GJ7	
5HA7	12FQ		
5HB7	9QA		
5HG8	9MP	5HG8/LCF86	
5HZ6	7EN	5GX6	
5J6	7BF	5MHH3	
5JK6	7CM	5JL6	
5JL6	7CM	5JK6	
5JW8	9AE		
5KD8	9AE		
5KE8	9DC		
5KZ8	9FZ	5CR8	5CM8
5LJ8	9GF	5MB8, 5GS7	
5MB8	9FA	5DH8, 5LJ8	5GS7, 5BE8, 5BR8, 5CL8, 5FV8
5MHH3	7BF	5J6	
5MQ8	9AE		5GH8, 5GH8A, 5CQ8
5R4	5T	5R4GB, 5R4GY, 5R4GYB	5T4, 5U4G, 5U4GB
5R4GB	5T	5R4, 5R4GY, 5R4GYB	5T4, 5U4G, 5U4GB
5R4GY	5T	5R4, 5R4GB, 5R4GYB	5T4, 5U4G, 5U4GB
5R4GYB	5T	5R4, 5R4GB 5R4GY	5T4, 5U4G, 5U4GB
5T4	5T	5R4, 5R4GB, 5R4GY, 5R4GYB, 5U4G, 5U4GB	

Original	Base Diagram	Preferred Substitution	Good Substitution
5T8	9E		
5T20		250TL	
5T21		250TH	
5T31		450TL	
5T31		450TH	
5T34		304TL	
5T35		304TH	
5U4	5T	5U4G, 5U4GB, 5AS4	5V4, 5V4A, 5AS4 5AU4
5U4G	5T	5U4, 5U4GB, 5AS4, 5R4, 5U4GB, 5R4GY, 5R4GYB	5V4, 5V4A, 5AU4
5U4GB	5T	5U4, 5U4G, 5AS4, 5R4, 5R4GB, 5R4GY, 5R4GYB	5V4, 5V4A, 5AU4
5U8	9AE	5GH8, 5GH8A, 5EA8	5CQ8
5U9	10K	5U9/LCF201	5X9
5V3	5T	5AU4, 5V3A	5AS4
5V3A	5T	5AU4, 5V3	5AS4
5V4	5L	5V4G, 5V4GA	5CZ4, 5Z4
5V4G	5L	5V4, 5V4GA	5CZ4, 5Z4
5V4GA	5L	5V4, 5V4G	5CZ4, 5Z4
5V6GT	7AC		
5W4	5L	5W3GT, 5AX4, 5AZ4, 5R4, 5R4GB, 5R4GY, 5R4GYB, 5T4	
5W4GT	5L	5W4, 5AX4, 5AZ4, 5R4, 5R4GB, 5R4GY, 5R4GYB, 5T4	
5X3	4C		
5X4G	5Q		

Original	Base Diagram	Preferred Substitution	Good Substitution
5X8	9AK		
5X9	10K		5U9
5Y3G	5T	5Y3GT, 5Y3GA, 5AX4, 5AZ4, 5T4, 5R4, 5R4GB, 5R4GY, 5R4GYB	5U4, 5U4G, 5U4GB
5Y3GA	5T	5Y3G, 5Y3GT, 5AX4, 5AZ4, 5T4, 5R4, 5R4GB, 5R4GY, 5R4GYB	5U4, 5U4G, 5U4GB
5Y3GT	5T	5Y3G, 5Y3GA, 5AX4, 5AZ4, 5TA, 5R4, 5R4GB, 5R4GY, 5R4GYB	5U4, 5U4G, 5U4GB
4l15Y4G	5Q	5Y4GT, 5Y4GA	
5Y4GA	5Q	5Y3G, 5Y4GT	
5Y4GT	5Q	5Y4G, 5Y4GA	
5Z3	4C	83	
5Z4	5L	5Y3G, 5Y3GA, 5Y3GT	5CG4, 5V4, 5V4G, 5V4GA
5-125B		4E27A	
5-500A			
6A3	4D		
6A6	7B		6E6
6A7	7C	6A7S	
6A7S	7C	6A7	
6A8	8A	6A8G, 6A8GT	6D8, 6D8G
6A8G	8A	6A8, 6A8GT	6D8, 6D8G
6A8GT	8A	6A8, 6A8G	6D8, 6D8G
6AB4	5CE	6664	6C4, 6DR4
6AB5	6R	6N5	6E5, 6U5, 6T5
6AB7	8N		6AC7, 6AJ7, 6SK7

Original	Base Diagram	Preferred Substitution	Good Substitution
6AC5GT	6Q		
6AC7	8N	6AC7W	
6AC7W	8N	6AC7	
6AC10	12FE		6D10, 6U10
6AD4	8DK	5898	6AK4, 5719
6AD6G	7AG		6AF6G
6AD7G	8AY		
6AD8	9HE		6DC8, 6N8, 6DC8/EBF89
6AD10	12EZ	6T10	
6AE5	6Q 6AF5, 6AE5GT		6C5, 6J5, 6C5GT, 5GT, 6L5, 6L5G, 6P5, 6P5GT
6AE5GT	6Q	6AF5, 6AE5	6C5, 6C5GT, 6J5, 6J5GT, 6L5, 6L5G, 6Pt, 6P5GT
6AE6G	7AH		
6AE7GT	7AX		
6AF3	9CB	6BR3	6AL3, 6A13/EY88, 6R
6AF4	7DK	6T4, 6AF4A	6AN4, 6DX4, 6DZ4
6AF4A	7DK	6AF4, 6T4	6AN4, 6DX4, 6DZ4
6AF5	6Q		6AE5, 6C5, 6AE5GT, 6C5GT, 6J5, 6J5GT, 6L5, 6L5G, 6P5, 6P5GT

41

Original	Base Diagram	Preferred Substitution	Good Substitution
6AF6G	7AG		6AD6G
6AF9	10L		
6AF11	12DP	6AS11	6BD11
6AG5	7BD	6BC5	6CE5
6AG7	8Y	6AK7. 6AG7Y	6L10
6AG7Y	8Y	6AG7. 6AK7	6L10
6AG9	12HE	6AL9	
6AG11	12DP		6AY11
6AH4GT	8EL		
6AH6	7BK	6HR6. 6HS6	6AU6. 6BA6. 6CB6
6AH9	12HJ		
6AJ4	9BX	6AM4	6CR4
6AJ5	7BD	6AK5. 6AK5/EF95	6AG5. 6BC5. 6CE5
6AJ7	8N	6AC7. 6AC7W	6AB7. 6SK7. 6SK7GT. 6SD7
6AJ8	6E	6AJ8/ECH81	
6AK4	8DK		5718. 5897. 6814. 8527
6AK5	7BD	5591. 5654. 6AK5/EF95	6AG5. 6BC5. 6CE5
6AK6	7BK		6AU6. 6BA6. 6HR6. 6HS6
6AK7	8Y	6AG7. 6AG7Y	
6AK8	9E	6AK8/EABC80. 6T8	6R8
6AK10	12FE		6AC10. 6D10. 6U10
6AL3	9CB	6AL3/EY88. 6BR3	6AF3. 6V3
6AL5	6BT	5726. 6663	6EB5
6AL6	6AM		6FH6. 6GW6
6AL7	8CH	6AL7GT	

Original	Base Diagram	Preferred Substitution	Good Substitution
6AL7GT	8CH	6AL7	
6AL11	12BU		6G11
6AM4	9BX	6AJ4	6CR4
6AM6	7BD	6AM6/EF91	
6AM8	9CY	6AM8A, 6HJ8	
6AM8A	9CY	6AM8, 6HJ8	
6AN4	7DK		6T4, 6AF4, 6AF4A
6AN5	7BD		6AK6
6AN8	9DA	6AN8A	
6AN8A	9DA	6AN8	
6AQ5	7BZ	6AQ5A, 6005, 6HG5, 6669	6HR5, 6DS5
6AQ5A	7BZ	6AQ5, 6005, 6HG5, 6669	6HR5, 6D55
6AQ6	7BT	6AT6, 6BT6	6AV6, 6BK6
6AQ7	8CK	6AQ7GT	
6AQ7GT	8CK	6AQ7	
6AQ8	9AJ	6AQ8/ECC85, 6BK7, 6DT8, 6EV7	6BC8, 6BQ7, 6BZ7, 6BZ8
6AR5	6CC		
6AR6	7BT		
6AR8	9DP	6JH8	
6AR11	12DM		
6AS4	4CG	6DM4, 6DQ4, 6DT4	6CQ4, 6DE4
6AS5	7CV	6CA5	6CU5
6AS6	7CM	6DB6, 5725	6GX6, 6GY6, 6HZ6
6AS7	8BD	6AS7G, 6080, 6AS7GA	6520, 5998, 7236, 7802
6AS7G	8BD	6AS7, 6080, 6AS7GA	6520, 5998, 7236, 7802

Original	Base Diagram	Preferred Substitution	Good Substitution
6AS7GA	8BD	6AS7, 6AS7G, 6080	6520, 5998, 7236, 7802
6AS8	9DS		
6AS11	12DP	6AF11	6BD11
6AT6	7BT	6BT6, 6AV6, 6BK6	6AQ6
6AT8	9DW	6AT8A	6CG8, 6BR8, 6FG7, 6FV8, 6JN8, 6MB8, 6LJ8
6AT8A	9DW	6AT8	6CG8, 6BR8, 6FG7, 6FV8, 6JN8, 6MB8, 6LJ8
6AU4	4CG	6AU4GT, 6AU4GTA, 6CQ4, 6DM4, 6DE4, 6DT4	6AX4, 6AX4GT, 6AX4GTB
6AU4GT	4CG	6AU4, 6AU4GTA, 6CG4, 6DM4, 6DE4, 6DT4	6AX4, 6AX4GT, 6AX4GTB
6AU4GTA	4CG	6AU4, 6AU4GT, 6CQ4, 6DM4, 6DE4, 6DT4	6AX4, 6AX4GT, 6AX4GTB
6AU5	6CK	6FW5, 6AU5GT	6AV5, 6AV5GT
6AU5GT	6CK	6AU5, 6FW5	6AV5, 6AV5GT
6AU6	7BK	6AU6A, 6136, 6HS6, 6BA6	6HS6, 6AH6, 6CG6, 6AG5, 6CE5, 6BC5
6AU6A	7BK	6AU6, 6136, 6HS6, 6BA6	6HS6, 6AH6, 6CG6, 6AG5, 6BC5, 6CE5

Original	Base Diagram	Preferred Substitution	Good Substitution
6AU7	9A	7AU7	6189
6AU8	9DX	6AU8A, 6BA8, 6BA8A, 6BH8	6AW8, 6AW8A, 6LF8, 6JT8, 6LB8
6AU8A	9DX	6AU8, 6BA8, 6BA8A, 6BH8	6AW8, 6AW8A, 6LF8, 6JT8, 6LB8
6AV5	6CK	6FW5, 6AV5GA, 6AV5GT	6AU5, 6AU5GT
6AV5GA	6CK	6AV5, 6FW5, 6AV5GT	6AU5, 6AU5GT
6AV5GT	6CK	6AV5, 6AV5GA, 6FW5	6AU5, 6AU5GT
6AV6	7BT	6BK6, 6AT6	6BT6, 6AQ6
6AV11	12BY	6AC10	6D10, 6K11
6AW6	7CM	6CB6	6CF6, 6DC6, 6DE6
6AW7	8CQ		
6AW8	9DX	6AW8A, 6AU8, 6AU8A	6BA8, 6BA8A, 6CX8, 6EB8, 6HF8, 6JE8, 6JT8, 6JV8, 6KS8, 6KV8, 6LY8, 6LF8
6AW8A	9DX	6AW8, 6AU8, 6AU8A	6BA8, 6BA8A, 6CX8, 6EB8, 6HF8, 6JE8, 6JT8,

(continued)

45

Original	Base Diagram	Preferred Substitution	Good Substitution
			6JV8, 6KS8, 6KV8, 6LY8, 6LF8
6AX3	12BL	6CD3, 6BE3, 6BZ3, 6CE3, 6DT3, 6CG3, 6BW3, 6BQ3	
6AX4	4CG	6AS4, 6DA4, 6DM4, 6DQ4, 6DT4 6DM4A, 6AX4GT, 6AX4GTB	6AU4, 6AU4GT, 6AU4GTA 6CQ4, 6DE4
6AX4GT	4CG	6AS4, 6DA4, 6DM4, 6DM4A, 6AX4GT 6AX4GTB	6AU4, 6AU4GT, 6AU4GTA, 6CQ4, 6DE4
6AX4GTB	4CG	6AX4, 6AX4GT, 6DM4, 6DA4, 6DM4A, 6DQ4, 6DT4	6AU4, 6AU4GT, 6AU4GTA, 6CQ4, 6DE4
6AX5	6S	6AX5GT	6W5
6AX5GT	6S	6AX5	6W5
6AX6	7Q		
6AX7	9A	6851	
6AX8	9AE		6EA8, 6GH8, 6GH8A, 6GJ8, 6U8, 6U8A, 6HL8, 6KD8, 6LM8, 6KE8, 6MG8
6AY3	9HP	6BH3, 6BH3A, 6AY3A, 6AY3B, 6DW4, 6DW4A, 6DW4B, 6CL3, 6CK3	

Original	Base Diagram	Preferred Substitution	Good Substitution
6AY3A	9HP	6AY3, 6BH3, 6BH3A, 6AY3B, 6DW4, 6DW4A, 6DW4B, 6CL3, 6CK3	
6AY3B	9HP	6AY3, 6AY3A, 6BH3, 6BH3A, 6DW4, 6DW4A, 6DW4B, 6CL3, 6CK3	
6AY11	12DA		6AG11
6AZ6	8EH	6184	
6AZ8	9ED		
6B3	9BD	6V3, 6V3A	6AF3, 6AL3, 6AL3/EY88, 6BR3, 6KR19, 6R3
6B4G	5S		
6B5	6AS		
6B6	7V	6Q7, 6B6G	6T7
6B6G	7V	6B6, 6Q7	6T7
6B7	7D	6B7S	
6B7S	7D	6B7	
6B8	8E	6B8G	
6B8G	8E	6B8	
6B10	12BF		
6BA3	9HP	6AY3, 6AY3A, 6AY3B, 6BH3, 6BH3A, 6CK3,	6DW4, 6DW4A, 6DW4B
6BA6	7BK	6CL3 6BA6 EF93, 6BA6W, 6660, 5749	6HR6, 6CG6, 6AU6, 6AU6A, 6HS6
6BA7	8CT		
6BA8	9DX	6BA8A, 6AW8, 6AW8A, 6BH8	6AU8, 6AU8A, 6JT8

Original	Base Diagram	Preferred Substitution	Good Substitution
6BA8A	9DX	6BA8, 6AW8, 6AW8A, 6BH8	6AU8, 6AU8A, 6JT8
6BA11	12ER		
6BC4	9DR		
6BC5	7BD	6CE5, 6189, 6AG5	6AH6, 6AU6, 6AU6A, 6HS6, 6CB6, 6CB6A, 6CF6
6BC7	9AX	6GQ7	
6BC8	9AJ	6BZ8, 6BQ7, 6BQ7A, 6BZ7	6AQ8, 6AQ8/ECC85, 6BK7
6BD4	8FU	6BD4A, 6BK4, 6BK4A, 6BK4B, 6BK4C, 6EL4A	
6BD4A	8FU	6BD4, 6BK4, 6BK4A, 6BK4B, 6BK4C, 6EL4A	
6BD5	6CK	6BD5GT	6AU5, 6AV5, 6FW5, 6AU5GT, 6AV5GT, 6AV5GA
6BD5GT	6CK	6BD5	6AU5, 6AV5, 6FW5, 6AU5GT, 6AV5GT, 6AU5GA
6BD6	7BK	6BA6, 6BA6/EF93, 5749, 5760	6HR6
6BD11	12DP	6AF11, 6AS11	
6BE3	12GA	6BZ3, 6CD3, 6DT3	6CE3, 6CG3, 6BW3, 6DQ3

Original	Base Diagram	Preferred Substitution	Good Substitution
6BE6	7CH	5750, 6BY6, 6BE6W	5915, 7036
6BE6W	7CH	6BE6, 5750, 6BY6	5915, 7036
6BE8	9EG	6BR8, 6BR8A, 6BE8A	6FV8, 6JN8, 8446, 6FV8A, 6CL8, 6CL8A, 6FG7
6BF5	7BZ		6DS5, 6HG5, 6HR5, 6AQ5, 6AQ5A
6BF6	7BT	6BU6	
6BF7	8DG	6BG7	6947, 7079, 7963
6BF11	12EZ		6AD10, 6T10
6BG6	6BT	6BGG, 6BG6GA	6GC6
6BG7	8DG	6BF7	6021, 8525
6BH3	9HP	6BH3A, 6AY3, 6AY3A, 6AY3B, 6CK3, 6CL3	6DW4, 6DW4A, 6DW4B, 6CH3, 6CJ3
6BH3A	9HP	6BH3, 6AY3, 6AY3A, 6AY3B, 6CK3, 6CL3	6DW4, 6DW4A, 6DW4B, 6CH3, 6CJ3
6BH6	7CM	6BJ6	6DB6
6BH8	9DX	6AU8, 6AU8A, 6AW8, 6AW8A	6DX8, 6JT8, 6KR8, 6BA8, 6BA8A
6BH11	12FP		
6BJ3	12BL	6AX3	6BZ3, 6CD3, 6BE3,

(continued)

49

Original	Base Diagram	Preferred Substitution	Good Substitution
			6CE3, 6CG3, 6DT3, 6DQ3, 6BW3
6BJ6	7CM	6DC6, 6BJ6A	6BH6, 6DB6
6BJ6A	7CM	6BJ6, 6DC6	6BH6, 6DB6
6BJ7	9AX		6BC7, 6GQ7
6BJ8	9ER		6BN8
6BK4	8GC	6BK4A, 6BK4B, 6BK4C, 6EL4, 6EL4A	6BD4, 6BD4A
6BK4A	8GC	6BK4, 6BK4B, 6BK4C, 6EL4, 6EL4A	6BD4, 6BD4A
6BK4B	8GC	6BK4, 6BK4A, 6BK4C, 6EL4, 6EL4A	6BD4, 6BD4A
6BK4C	8GC	6BK4, 6BK4A, 6BK4B, 6EL4, 6EL4A	6BD4, 6BD4A
6BK5	9HQ		
6BK6	7BT	6AV6	6AT6, 6BT6, 6AQ6
6BK7	9AJ	6FW8, 6KN8, 6BK7A, 6BK7B	6AQ8, 6BC8, 6BQ7, 6BQ7A, 6BZ7, 6BS8, 6BZ8, 6CH7
6BK7A	9AJ	6BK7, 6BK7B, 6FW8, 6KN8	6AQ8, 6BC8, 6BQ7, 6BQ7A, 6BZ7, 6BS8, 6BZ8, 6CH7
6BK7B	9AJ	6BK7, 6BK7A, 6FW8, 6KN8	6AQ8, 6BC8, 6BQ7,

Original	Base Diagram	Preferred Substitution	Good Substitution
			6BQ7A, 6BZ7, 6BS8, 6BZ8, 6CH7
6BK11	12BY	6K11, 6AC10	6BQ11
6BL4	8GB	6DM4, 6DM4A, 6DA4, 6DQ4, 6DE4, 6CQ4	6DT4, 6AU4, 6AU4GT, 6AU4GTA
6BL7	8BD	6BX7, 6BL7GT, 6BI7GTA	
6BL7GT	8BD	6BL7, 6BX7, 6BL7GTA	
6BL7GTA	8BD	6BL7, 6BL7GT, 6BX7	
6BL8	9AE	6BL8/ECF80, 6LN8	6AX8, 6U8, 6U8A, 6KE8, 6MG8
6BM5	7BZ	6AQ5, 6AQ5A, 6HG5	6DL5, 6DS5
6BM8	9EX	6BM8/ECL82	6HC8
6BN4	7EG	6BN4A	
6BN4A	7EG	6BN4	
6BN6	7DF	6KS6	
6BN8	9ER		6BJ8
6BN11	12GF		
6BQ5	9CV	6BQ5/EL84, 7189	
6BQ6	6AM	6BQ6GT, 6BQ6GTB, 6CU6	6DQ6, 6FH6, 6GW6
6BQ6GT	6AM	6BQ6, 6BQ6GTB, 6CU6	6DQ6, 6FH6, 6GW6
6BQ6GTB	6AM	6BQ6, 6BQ6GT, 6CU6	6DQ6, 6FH6, 6GW6
6BQ7	9AJ	6BQ7A, 6BZ7, 6BS8	6BC8, 6BK7, 6BK7A, 6BK7B, 6CH7, 6BZ8

Original	Base Diagram	Preferred Substitution	Good Substitution
6BQ7A	9AJ	6BQ7, 6BZ7, 6BS8	6BC8, 6BK7, 6BK7A, 6BK7B, 6CH7, 6BZ8
6BR3	9CB	6RK19	6AF3, 6V3, 6V3A, 6AL3
6BR5	9DB	6DA5	
6BR8	9FA	6BR8A, 6FV8, 6FV8A, 6JN8	6BE8, 6CL8, 6CL8A, 6FG7
6BR8A	9FA	6BR8, 6FV8, 6FV8A, 6JN8	6BE8, 6CL8, 6CL8A, 6FG7
6BS3	9HP	6BS3A, 6CK3, 6CL3	6CH3, 6CJ3
6BR8A	9FA	6BR8, 6FV8, 6FV8A, 6JN8	6BE8, 6CL8, 6CL8A, 6FG7
6BS3	9HP	6BS3A, 6CK3, 6CL3	6CH3, 6CJ3
6BS3A	9HP	6BS3, 6CK3, 6CL3	6CH3, 6CJ3
6BS8	9AJ	6BC8, 6BQ7, 6BQ7A, 6BZ7	6BX8, 6BZ8, 6JK8, 6CH7, 6CX7
6BT6	7BT	6AT6	6AV6, 6BK6, 6AQ6
6BU4	8GC	6BK4, 6BK4A, 6BK4B, 6BK4C, 6EL4, 6EL4A	
6BU6	7BT	6BF6	
6BU8	9FG	6HS8, 6GS8, 6KF8	
6BV8	9FJ		
6BV11	12HB		
6BW3	12FX	6CG3, 6DQ3,	6AX3, 6BE3
6BW4	9DJ		

Original	Base Diagram	Preferred Substitution	Good Substitution
6BW6	9AM		6061
6BW7	9AQ	6BX6, 6HM6	6EL7, 6EJ7, 6EJ7/EF184 6JC6, 6JC6A
6BW8	9HK		
6BW11	12HD		
6BX6	9AQ	6BW7, 6HM6	6EL7, 6EJ7, 6EJ7./EF184, 6JC6, 6JC6A
6BX7	8BD	6BL7, 6BL7GT, 6BL7GTA, 6BX7GT	6DN7
6BX7GT GT	8BD	6BX7, 6BL7,	6DN7 6BL7GT
6BX8	9AJ	6DJ8, 6DJ8/ECC88, 6ES8	6BC8, 6BQ7, 6BQ7A, 6BS8, 6BZ7, 6BZ8
6BY5GA	6CN		
6BY6	7CH		6CS6, 6BE6, 5750, 5915, 7036
6BY7	9AQ	6BX6	6EC7
6BY8	9FN		
6BY11	12EZ		6AD10, 6T10
6BZ3	12FX	6CD3, 6BE3	6CE3, 6CG3
6BZ6	.7CM	6GM6, 6HQ6, 6JH6	6JK6
6BZ7	9AJ	6BQ7, 6BQ7A, 6BS8, 6BC8	6BX8, 6BZ8, 6CH7, 6CX7
6BZ8	9AJ	6BC8	6BQ7, 6BQ7A, 6BZ7, 6CH7, 6CX7

Original	Base Diagram	Preferred Substitution	Good Substitution
6C4	6BG	6C4W, 6C4WA, 6135	6AB4, 6DR4
6C4W	6BG	6CW4, 6C4WA, 6135	6AB4, 6DR4
6C4WA	6BG	6C4, 6C4W, 6135	6AB4, 6DR4
6C5	6Q	6C5G, 6C5GT, 6J5, 6J5GT	6AE5, 6L5, 6L5G, 6P5, 6P5GT
6C5G	6Q	6C5, 6C5GT, 6J5, 6J5GT	6AE5, 6L5, 6L5G, 6P5, 6P5GT
6C5GT	6Q	6C5, 6C5G, 6J5, 6J5GT	6AE5, 6L5, 6L5G, 6P5, 6P5GT
6C6	6F	77	
6C7	7G		
6C8	8G	6C8G, 6F8, 6F8G	
6C8G	8G	6C8, 6F8, 6F8G	
6C9	12F		
6C10	12BQ	6BK11	6K11, 6Q11
6C16	9AE	6BL8 ʳB ₊8/ECʳ80	
6C21			
6C31	8K	6K8, 6K8G, 6K8GT	
6CA4	9M		
6CA5	7CV		6CU5, 6AS5, 6EH5
6CA7	8ET	6CA7/EL34	
6CB5	8GD	6CB5A, 6CL5	
6CB6	7CM	6CB6A, 6CF6, 6BZ6, 6DK6, 6DE6, 6JH6	6HQ6, 6JK6, 6JL6
6CB6A	7CM	6CB6, 6CF6, 6BZ6, 6DK6, 6DE6, 6JH6	6HQ6, 6JK6, 6JL6
6CC10	8BD	5692	

54

Original	Base Diagram	Preferred Substitution	Good Substitution
6CC31	7BF	6J6, 6J6A, 6J6WA, 6J6WB	
6CC43	9AJ	6AQ8, 6AQ8/ECC85	
6CD3	12FX	6CE3, 6DT3	6CG3
6CD6	5BT	6CD6G, 6CD6GA, 6EX6	6DN6, 6DQ5, 6GC6
6CD6G	5BT	6CD6, 6CD6GA, 6EX6	6DN6, 6DQ6, 6GC6
6CD6G	5BT	6CD6, 6CD6GA, 6EX6	6DN6, 6DQ5, 6GC6
6CD6GA	5BT	6CD6, 6CD6G, 6EX6	6DN6, 6DQ5, 6GC6
6CE3	12GK	6CD3, 6DT3	6CG3
6CE5	7BD	6BC5	6AG5, 6CB6, 6CB6A, 6CF6, 6DK6
6CF6	7CM	6CB6, 6CB6A, 6BZ6, 6DE6, 6DK6	6HQ6, 6JH6, 6JK6
6CG3	12HF	6BW3, 6DQ3	6CD3, 6CE3, 6DT3
6CG6	7BK	6BA6, 6BA6/EF93, 6BD6	6AU6, 6AU6A, 6HR6, 6HS6
6CG7	9AJ	6FQ7	6GU7
6CG8	9GF	6FG7, 6LJ8, 6CG8A	6AT8, 6BE8, 6BR8, 6AT8A, 6BR8A, 6CL8, 6CL8A, 6FV8, 6FV8A, 6JN8, 6MB8

Original	Base Diagram	Preferred Substitution	Good Substitution
6CG8A	9GF	6CG8, 6FG7, 6LJ8	6AT8, 6AT8A, 6BE8, 6BR8, 6BR8A, 6CL8, 6CL8A, 6FV8, 6FV8A, 6JN8, 6MB8
6CH3	9HP	6CJ3	
6CH7	9FC	6CX7	6BC8, 6BQ7, 6BQ7A, 6BS8, 6BZ7
6CH8	9FT		6CU8
6CJ3	9HP	6CH3	
6CJ6		6DR6	
6CK3	9HP	6CL3, 6CJ3, 6CH3	
6CK4	8JB		6AH4
6CL3	9HP	6CK3, 6CJ3, 6CH3	
6CL5	8GD	6CB5	
6CL6	9BV	6677, 6197	
6CL8	9FX	6CL8A, 6BR8, 6BR8A	6FV8, 6FV8A, 6JN8
6CL8A	9FX	6CL8, 6BR8, 6BR8A	6FV8, 6FV8A, 6JN8
6CM3	9HP	6CJ3, 6CH3	
6CM6	9CK	6CZ5	6DT5, 6CS5
6CM7	9ES		
6CM8	9FZ		
6CN7	9EN		
6CQ4	4CG	6DE4	6DM4, 6DT4
6CQ6	7BK		6065

Original	Base Diagram	Preferred Substitution	Good Substitution
6CQ8	9GE		6EA8, 6GH8, 6GH8A, 6LM8, 6U8, 6U8A
6CR4	9BX		6AJ4, 6AM4
6CR6	7EA		
6CR8	9GJ		6CS8, 6KZ8, 6CM8
6CS5	9GR	6DB5	6GC5, 6CM6
6CS6	7CH		6BY6, 6BE6
6CS7	9EF		6DE7
6CS8	9FZ	6CM8, 6KZ8	6CR8
6CT3	9RX		
6CU5	7CV	6CA5	6AS5, 6EH5
6CU6	6AM	6BQ6, 6BQ6GT, 6BQ6GTB, 6DQ6, 6DQ6A, 6DQ6B, 6GW6	
6CU8	9GM		6CH8
6CW4	12AQ	6DS4	
6CW5	9CV	6CW5/EL86	6HL5
6CW7	9FC	6FC7	
6CX7	9FC	6CH7	6BC8, 6BQ7, 6BQ7A, 6BS8, 6BZ7
6CX8	9DX	6AU8, 6AU8A, 6BH8	6JT8, 6KR8, 6LB8
6CY5	7EW	6FV6	6AE5, 6EV5
6CY7	9LG		6DR7, 6FD7, 6FR7
6CZ5	9HN	6CM6	6DT5, 6CS5
6D2	6BT	6AL5	5726
6D4	5AY		
6D6	6F	78	

Original	Base Diagram	Preferred Substitution	Good Substitution
6D8	8A	6A8, 6A8G, 6A8GT, 6D8G	
6D8G	8A	6D8, 6A8, 6A8G, 6A8GT	
6D10	12BY	6AC10	6AV11
6DA4	4CG	6DM4, 6DQ4, 6DT4, 6DM4A	6AU4, 6AU4GT, 6AU4GTA, 6CQ4, 6DE4
6DA5	9DB	6BR5	
6DA7	9EF		6DE7, 6EW7
6DB5	9GR	6CS5	6CM6, 6GC5
6DB6	7CM		6AS6, 6BJ6
6DC6	7CM	6HQ6, 6JH6	6BJ6, 6GM6
6DC8	9HE	6DC8/EBF89	6AD8, 6N8
6DE4	4CG	6CQ4, 6DT4	6DM4
6DE6	7CM	6CB6, 6CB6A, 6CF6, 6DK6	6BZ6, 6HQ6
6DE7	9HF	6EW7	6DA7
6DG6	7S	6DG6GT, 6W6, 6W6GT	6F6, 6F6G, 6F6GT, 6V6, 6V6GT, 6V6GTA, 6V6GTY, 6U6
6DG6GT	7S	6DG6, 6W6, 6W6GT	6F6, 6F6G, 6F6GT, 6V6, 6V6GT, 6V6GTA, 6V6GTY, 6U6
6DJ8	9AJ	6DJ8/ECC88, 6ES8, 6FW8, 6ES8/ECC189	6BK7, 6ES8, 6FW8, 6BK7B
6DK3	9SG	6DL3	

Original	Base Diagram	Preferred Substitution	Good Substitution
6DK6	7CM	6CB6, 6CB6A, 6CF6, 6DE6, 6JK6	6BZ6, 6HQ6
6DL3	9GD	6DK3	
6DL4	9NY	6DL4/EC88, 8255	
6DL5	7BZ	6DL5/EL95	6BM5
6DM4	4CG	6DM4A, 6DA4, 6DT4	6CQ4, 6DE4
6DM4A	4CG	6DM4, 6DA4, 6DT4	6CQ4, 6DE4
6DN3	9HP	6CH3, 6CJ3	
6DN6	6BT	6CD6	6EX6
6DN7	8BD		6EA7, 6EM7
6DQ3	12HF	6CG3, 6BW3	
6DQ4	4CG	6DM4, 6DM4A, 6DA4, 6DT4	6CQ4, 6DE4, 6AX4, 6AX4GT, 6AX5GTB
6DQ5	8JC		6CD6, 6EX6
6DQ6	6AM	6DQ6A, 6DQ6B, 6GW6, 6FH6	
6DQ6A	6AM	6DQ6, 6DQ6B, 6GW6, 6FH6	
6DQ6B	6AM	6DQ6, 6DQ6A, 6GW6, 6FH6	
6DR6		6CJ6	
6DR7	9HF	6FD7	6CY7, 6FR7
6DS4	12AQ	6CW4	
6DS5	7BZ		6AQ5, 6AQ5A, 6HG5, 6HR5
6DT3	12GK	6CE3, 6CD3	6CG3
6DT4	4CG	6DA4, 6DM4, 6DM4A, 6DQ4	6CQ4, 6DE4, 6AU4, 6AU4GT, 6AU4GTA
6DT5	9HN		6CM6, 6CZ5

Original	Base Diagram	Preferred Substitution	Good Substitution
6DT6	7EN	6DT6A	
6DT6A	7EN	6DT6	
6DT8	9AJ		6AQ8, 6AQ8/ECC85, 6BZ8, 6CH7, 6CX7, 6EV7
6DV4	12EA		
6DW4	9HP	6DW4A, 6DW4B, 6CL3	6CH3, 6CK3, 6CJ3
6DW4A	9HP	6DW4, 6DW4B, 6CL3	6CH3, 6CK3, 6CJ3
6DW4B	9HP	6DW4, 6DW4A, 6CL3	6CH3, 6CJ3, 6CK3
6DW5	9CK	6CM6	6DB5, 6CZ5, 6DT5
6DX4	7DK	6DY4	6AF4, 6AF4A, 6T4, 6AN4, 6DZ4
6DX8	9HX	6DX8/ECL84	
6DY4	7DK	6DX4	6AF4, 6AF4A, 6T4, 6AN4, 6DZ4
6DY5	9CV	6BQ5, 6BQ5/EL84	6CW5
6CY7	8JP		6DZ7
6DZ4	7DK	6AF4, 6AF4A, 6T4	6AN4, 6DX4, 6DY4
6DZ7	8JP		6DY7
6DZ8	9JE		6FY8
6E5	6R	6U5	6T5, 6AB5, 6N5
6E6	7B		6A6
6E7	7H		6D7

Original	Base Diagram	Preferred Substitution	Good Substitution
6EA4	12FA	6EH4, 6EH4A	
6EA5	7EW	6EV5, 6AK5, 6AK5/EF95, 6CY5	6FV6
6EA7	8BD	6GL7, 6EM7	6DN7
6EA8	9AE	6GH8, 6GH8A, 6GJ8	6HL8, 6KD8, 6LM8, 6U8, 6U8A
6EB5	6BT	6AL5	6D2, 5726
6EB8	9DX	6GN8, 6JT8, 6LY8	6AU8, 6AU8A, 6AW8, 6AW8A, 6CX8, 6HF8, 6HZ8, 6JE8, 6JV8, 6KS8, 6KV8, 6KV8A, 6LF8
6EC4	6EC4	6EC4A, 6EC4A/EY500	
6EC4A	6EC4	6EC4, 6EC4A/EY500	
6EF4	12HC	6EJ4, 6EJ4A	
6EF6	7S	6EY6, 6EZ5	6W6, 6W6GT, 6DG6, 6DG6GT
6EH4	12FA	6EH4A, 6EA4	
6EH4A	12FA	6EH4, 6EA4	
6EH5	7CV	6GZ5	6AS5, 6CU5, 6CA5
6EH7	9AQ	6EH7/EF183	6EJ7, 6EJ7/EF184
6EH8	9JG	6AU8, 6AU8A	6AW8, 6AW8A

Original	Base Diagram	Preferred Substitution	Good Substitution
6EJ4	12HC	6EJ4A, 6EF4	
6EJ4A	12HC	6EJ4, 6EF4	
6EJ7	9AQ	6EJ7/EF184	6EH7, 6EH7/EF183, 6GK7
6EL4	8GC	6EL4A, 6BK4, 6BK4A, 6BK4B, 6BK4C	
6EL4A	8GC	6EL4, 6BK4, 6BK4A, 6BK4B, 6BK4C	
6EL7	9AQ	6BW7, 6HM6, 6JC6, 6JC6A, 6EJ7, 6EJ7/EF184	6BX6
6EM5	9HN		6CZ5, 6DT5, 6CM6
6EM7	8BD	6EA7	6DN7, 6GL7
6EN4	8NJ		
6EQ7	9LQ	6KL8	
6ER6	7FP	6ES5	6FY5, 6FQ5, 6GK5
6ES5	7FP	6ER5	6FY5, 6FQ5, 6GK5
6ES6	7BK	6ET6	
6ES8	9AJ	6ES8/ECC189, 6DJ8, 6DJ8/ECC88	6DT8
6ET6	7BK	6ES6	6FD6
6ET7	9LT	6KU8	
6EU7	9LS		
6EU8	9JF		
6EV5	7EW	6EA5	6FV6
6EV7	9LP		6FQ7, 6CG7, 6GU7
6EW6	7CM	6GM6	6BJ6, 6BX6, 6CB6, 6CB6A

62

Original	Base Diagram	Preferred Substitution	Good Substitution
6EW7	9HF	6DE7	6DR7, 6FD7, 6FR7
6EX6	5BT	6CD6, 6CD6GA	
6EY6	7AC	6EZ5	
6EZ5	7AC	6EY6	
6EZ8	9KA		
6F4	7BR		6L4
6F5	5M	6F5G, 6F5GT, 6F5MG	
6F5G	5M	6F5, 6F5GT, 6F5MG	
6F5GT	5M	6F5, 6F5G, 6F5MG	
6F5MG	5M	6F5, 6F5G, 6F5GT	
6F6	7S	6F6G, 6F6GT, 6F6MG	
6F6G	7S	6F6, 6F6GT, 6F6MG	
6F6GT	7S	6F6, 6F6G, 6F6MG	
6F6MG	7S	6F6, 6F6G, 6F6GT	
6F7	7E		
6F8	8G	6F8G, 6C8, 6C8G	
6F8G	8G	6F8, 6C8, 6C8G	
6F10	8N	6AC7, 6AC7W	
6F24	9AQ	6EJ7, 6EJ7/EF184	6F30
6F29	9AQ	6EH7, 6EJ7/EF183	
6F30	9AQ	6EJ7, 6EJ7/EF184	6F24
6F31	7BK	6AU6, 6AU6A	
6F32	7BD	6AK5, 6AK5/EF95	
6F36	7BK	6AH6	
6F50R		4X500A	
6F50RA		4X500A	
6FA7	9MR		

Original	Base Diagram	Preferred Substitution	Good Substitution
6FC7	9DD	6CW7	
6FD6	7BK		6ES6, 6ET6
6FD7	9HF	6DE7, 6EW7, 6FR7	6DR7
6FE5	8KB		
6FG5	7GA	6FS5	6GU5, 6HS6
6FG6	9GA	6FG6G, 6FG6/EM84	6HU6
6FG7	9GF	6CL8, 6CL8A	6BE8, 6BR8, 6BR8A
6FH5	7FP	6ER5, 6ES5, 6FQ5, 6FY5, 6GK5	
6FH6	6AM	6DQ6, 6DQ6A, 6DQ6B, 6GW6	6BQ6, 6BQ6GT, 6BQ6GTB
6FH8	9KP		
6FJ7	12BM		
6FM7	12EJ		
6FM8	9KR		
6FQ5	7FP	6FQ5A, 6GK5	
6FQ5A	7FP	6FQ5, 6GK5	
6FQ7	9LP	6CG7	6EV7, 6GU7
6FR7	9HF	6EW7, 6FD7	6DE7, 6DR7
6FS5	7GA	6GU5	6FG5
6FV6	7FQ	6EV5	6CY5, 6EA5
6FV8	9FA	6FV8A, 6BR8, 6BR8A	6JN8, 6FG7, 6MB8
6FV8A	9FA	6FV8, 6BR8, 6BR8A	6JN8, 6FG7, 6MB8
6FW5	6CK		6AU5, 6AV5, 6AU5GT, 6AV5GA, 6AV5GT
6FW8	9AJ	6BC8, 6KN8, 6RHH8, 6DJ8, 6DJ8/ECC88, 6ES8,	6BK7, 6BK7A, 6BK7B, 6AW8,

Original	Base Diagram	Preferred Substitution	Good Substitution
		6ES8/ECC189.	6AW8A, 6BZ7
6FY5	7FP	6FY5/EC97, 6ES5, 6GK5	6ER5, 6FQ5, 6FQ5A
6FY7	12EO		
6FY8	9EX		6BM8, 6BM8/ECL82
6G5	6R	6H5, 6U5	6E5, 6T5
6G6G	7S		
6G11	12BU		6AL11
6GB3A	6AM	6BQ6, 6BQ6GT, 6BQ6GTB, 6CU6	
6GB5	9NH	6GB5/EL500	
6GB6	6AM	6DQ6, 6DQ6A, 6DQ6B, 6GW6, 6GB7, 6GB9	
6GB7	6AM	6DQ6, 6DQ6A, 6DQ6B, 6GW6, 6GB6, 6GB9	
6GB9	6AM	6DQ6, 6DQ6A, 6DQ6B, 6GW6, 6GB6, 6GB7	
6GC5	9EU		
6GD7	9GF	6LJ8, 6CG8, 6CG8A, 6FG7	6MB8
6GE5	12BJ		
6GE8	9LC	7734	
6GF5	12BJ		6GE5
6GF7	9QD	6GF7A	
6GR7A	9QD	6GF7	
6GH8	9AE	6GH8A, 6EA8, 6GJ8, 6LM8	6AX8, 6HL8, 6KD8, 6U8, 6U8A
6GH8A	9AE	6GH8, 6EA8, 6GJ8, 6LM8	6AX8, 6HL8, 6KD8, 6U8, 6U8A

(continued)

Original	Base Diagram	Preferred Substitution	Good Substitution
6GJ5	9QK	6JB6, 6JB6A, 6GJ5A	
6GJ5A	9QK	6JB6, 6JB6A, 6GJ5	
6GJ7	9QA	6GJ7/ECF801	6GX7, 6HB7, 6HD7, 6HJ7
6GJ8	9AE	6EA8, 6GH8, 6GH8A, 6HL8, 6LM8, 6MU8	6AX8, 6KD8, 6U8, 6U8A
6GK5	7FP	6FQ5, 6FQ5A, 6ER5	6ES5, 6FH5, 6FY5
6GK6	9GK		6HB6
6GK7	9AQ		6EH7, 6EH7/EF184, 6EJ7, 6EJ7/EF183
6GK17	4CG	6AU4, 6AU5GT, 6AU4GTA	
6GL7	8BD	6EA7, 6EM7	6DN7
6GM5	9MQ		
6GM6	7CM	6EW6	6BH6, 6BJ6, 6BZ6, 6CB6, 6CB6A, 6CF6, 6DC6, 6DE6, 6DK6, 6HQ6, 6JH6, 6JK6, 6JL6
6GM8	9DE		6DJ8, 6DJ8/ECC88, 6ES8
6GN8	9DX	6EB8, 6LY8, 6JT8	6AW8, 6AW8A, 6CX6,

Original	Base Diagram	Preferred Substitution	Good Substitution
			6HF8, 6HZ8, 6JE8, 6JV8, 6KR8, 6KV8, 6LB8, 6LF8, 6LQ8
6GQ7	9QM	6BC7, 6BJ7	
6GS7	9GF		6LJ8, 6FG7
6GS8	9FG	6BU8, 6HS8, 6MK8, 6MK8A	6KF8
6GU5	7GA	6FS5	
6GU7	9LP	6FQ7, 6CG7	6EV7
6GV5	12DR	6GY5	
6GV8	9LY	6GV8/ECL85	
6GW6	7EN	6GY6	6DT6, 6DT6A, 6HZ6
6GX7	9QA	6HB7, 6HD7, 6HJ7, 6GJ7	
6GY5	12DR	6GV5	
6GY6	7EN	6GX6	6HZ6
6GY8	9MB		
6GZ5	7CY		
6H5	6R	6G5, 6U5	6E5, 6T5
6H6	7Q	6H6G, 6H6GT, 6H6MG	
6H6G	7Q	6H6, 6H6GT, 6H6MG	
6H6GT	7Q	6H6, 6H6G, 6H6MG	
6H6MG	7Q	6H6, 6H6G, 6H6GT	
6H31	7CH	6BE6	
6HA5	7GM	6HA5S, 6HM5, 6HK5	6HQ5

Original	Base Diagram	Preferred Substitution	Good Substitution
6HA5S	7GM	6HA5, 6HM5, 6HK5	6HQ5
6HA6	9NW		6HB6
6HB5	12BJ		6GE5
6HB6	9NW	6HA6	
6HB7	9QA		6HD7, 6HJ7
6HC8	9EX	6BM8, 6BM8/ECL82	
6HD7	9QA	6GX7, 6HJ7	6HB7
6HE5	12EY	6JB5	
6HF5	12FB		6GV5
6HF8	9DX	6EB8, 6GN8	6HF8, 6HZ8, 6LY8, 6JE8, 6JT8, 6JV8, 6KR8, 6KV8, 6LQ8
6HG5	7BZ	6AQ5, 6AQ5A	6HR5
6HG8	9MP	6HG8/ECF86	
6HJ5	12FL		
6HJ7	9QA	6HD7, 6GX7	6HB7
6HJ8	9CY	6AM8, 6AM8A	
6HK5	7GM	6HA5, 6HA5S, 6HM5, 6HQ5	
6HK8	9AJ	6BC8, 6BZ8, 6BQ7, 6BQ7A, 6BZ7	6BS8, 6BK7, 6BK7A, 6BK7B
6HL8	9AE	6EA8, 6GH8, 6GH8A	6BL8, 6BL8 ECF80, 6BL8, 6KD8, 6LM8, 6U8, 6U8A, 6GJ8
6HM5	7GM	6HA5, 6HA5S, 6HK5	6HQ5
6HM6	9PM	6HT6, 6JC6, 6JC6A	6KT6

68

Original	Base Diagram	Preferred Substitution	Good Substitution
6HQ5	7GM	6HA5, 6HA5S, 6HM5, 6HK5	
6HQ6	7CM	6BZ6, 6JH6, 6GM6	6CB6, 6CB6A, 6DK6, 6CF6, 6DE6, 6JK6
6HR5	7BZ		6AQ5, 6AQ5A, 6HG5
6HR6	7BK	6HS6	6AU6, 6AU6A
6HS5	12GY	6HV5, 6HV5A	
6HS6	12GY	6HV5, 6HV5A	
6HS6	7BK	6HR6	6AU6, 6AU6A
6HS8	9GF	6BU8, 6GS8, 6KF8	
6HT6	9PM	6HM6, 6JC6, 6JC6A	6KT6, 6JD6
6HU6	9AG	6HU6/EM87	6FG6
6HU8	9NJ	6HU8/ELL80	
6HV5	12GY	6HV5A, 6HS5	
6HZ5	12JE	6JH5, 6JD5	
6HZ6	7EN	6GX6, 6GY6	6DT6
6HZ8	9DX	6AW8, 6AW8A, 6HF8, 6LF8, 6JV8	6JE8, 6CX8, 6EB8, 6GN8, 6HF8, 6KV8
6J4	7BQ	6J4WA, 8532	7137, 7245
6J4WA	7BQ	6J4, 8532	7137, 7245
6J5	6Q	6J5G, 6J5GT, 6J5GX, 6J5GTX, 6J5MG, 6C5, 6C5G, 6C5GT	6AE5, 6L5, 6L5G
6J5G	6Q	6J5, 6J5GT, 6J5GX, 6J5GTX, 6J5MG, 6C5, 6C5G, 6C5GT	6AE5, 6L5, 6L5G

Original	Base Diagram	Preferred Substitution	Good Substitution
6J5GT	6Q	6J5, 6J5G, 6J5GTX, 6J5GX, 6J5MG, 6C5, 6C5GT	6AE5, 6L5, 6L5G
6J5GTX	6Q	6J5, 6J5G, 6J5GT, 6J5GX, 6J5MG, 6C5, 6C5GT	6AE5, 6L5, 6L5G
6J5GX	6Q	6J5, 6J5G, 6J5GT, 6J5GTX, 6J5MG, 6C5, 6C5GT	6AE5, 6L5, 6L5G
6J5MG	6Q	6J5, 6J5G, 6J5GT, 6J5GTX, 6J5GX, 6C5, 6C5GT	6AE5, 6L5, 6L5G
6J6	7BF	6J6A, 7244, 6J6WA, 6J6WB	5964, 6045, 6101
6J6A	7BF	6J6, 6J6WA, 6J6WB, 7244	5964, 6045, 6101
6J6WA	7BF	6J6, 6J6A, 6J6WB, 7244	5964, 6045, 6101
6J6WB	7BF	6J6, 6J6A, 6J6WA, 7244	6964, 6045, 6101
6J7	7R	6J7G, 6J7GT, 6J7GTX, 6J7MG, 6W7, 6W7G	6K7, 6K7G, 6K7GT, 6K7GTX, 6K7MG, 6S7, 6S7G, 6U7, 6U7G
6J7G	7R	6J7, 6J7GT, 6J7GTX, 6J7MG 6W7, 6W7G	6K7, 6K7G, 6K7GT, 6K7GTX, 6K7MG, 6S7, 6S7G, 6U7, 6U7G
6J7GT	7R	6J7, 6J7G, 6J7GTX, 6J7MG, 6W7, 6W7G	6K7, 6K7G, 6K7GT, 6K7GTX, 6K7MG,

Original	Base Diagram	Preferred Substitution	Good Substitution
			6S7, 6S7G, 6U7, 6U7G
6J7GTX	7R	6J7, 6J7G, 6J7GT, 6J7MG, 6W7, 6W7G	6K7, 6K7G, 6K7GT, 6K7GTX, 6K7MG, 6S7, 6S7G, 6U7, 6U7G
6J7MG	7R	6J7, 6J7G, 6J7GT, 6J7GTX, 6W7, 6W7G	6K7, 6K7G, 6K7GT, 6K7GTX, 6K7MG, 6S7, 6S7G, 6U7, 6U7G
6J8	8H	6J8G	
6J8G	8H	6J8	
6J9	12G		
6J10	12BT	6Z10	
6J11	12BW		
6JA5	12EY		
6JA8	9DX	6CX8, 6LF8	6LQ8, 6LY8, 6EB8, 6GN8, 6HF8, 6JE8, 6JT8, 6JV8, 6KR8, 6KS8, 6KV8
6JB5	12EY	6HE5, 6JC5	
6JB6	9QL	6JB6A	6KM6, 6JU6
6JB6A	9QL	6JB6	6KM6, 6JU6
6JC5	12EY	6HE5, 6JB5	
6JC6	9PM	6JC6A, 6HM6	6HT6, 6JD6, 6KT6
6JC6A	9PM	6JC6, 6HM6	6HT6, 6JD6, 6KT6
6JC8	9PA		

Original	Base Diagram	Preferred Substitution	Good Substitution
6JD5	12JE	6HZ5, 6JH5	
6JD6	9PM	6JC6, 6JC6A, 6KT6	6HM6, 6HT6
6JE6	9QL	6JE6A, 6JE6B, 6JE6C, 6LQ6, 6MJ6	
6JE6A	9QL	6JE6, 6JE6B, 6JE6C, 6LQ6, 6MJ6	
6JE6B	9QL	6JE6, 6JE6A, 6JE6B, 6JE6C, 6LQ6, 6MJ6	
6JE6C	9QL	6JE6, 6JE6A, 6JE6B, 6LQ6, 6MJ6	
6JE8	9DX	6HF8, 6HZ8	6CX8, 6EB8, 6GN8, 6JT8, 6JV8, 6KV8, 6LY8
6JF6	9QL	6JU6, 6KM6	
6JG5	9SF		
6JG6	9QU	6JG6A, 6KV6	
6JG6A	9QU	6JG6, 6KV6	
6JH5	12JE	6HZ5, 6JD5	
6JH6	7CM	6HQ6	6EW6, 6BZ6
6JH8	9DP	6AR8	
6JK6	7CM	6JL6, 6EW6, 6GM6	6HQ6
6JK8	9AJ	6FW8, 6KN8	6BC8, 6BQ7, 6BQ7A, 6BZ7
6JL6	7CM	6JK6, 6EW6, 6GM6	6HQ6
6JL8	9DX	6CX8	6EB8, 6GN8, 6JE8, 6KR8, 6LB8, 6LY8

72

Original	Base Diagram	Preferred Substitution	Good Substitution
6JM6	12FJ	6JM6A	
6JM6A	12FJ	6JM6	
6JN6	12FK	6JN6A	
6JN6A	12FK	6JN6	
6JN8	9FA	6BR8, 6BR8A, 6KV8	6CL8, 6CL8A, 6FG7
6JQ6	9RA		
6JR6	9QU	6JG6, 6KV6, 6JG6A	
6JS6	12FY	6JS6A, 6JS6B, 6JS6C	6KD6, 6LB6, 6LF6, 6LX6, 6LR6
6JS6A	12FY	6JS6, 6JS6B, 6JS6C	6KD6, 6LB6, 6LF6, 6LX6, 6LR6
6JS6B	12FY	6JS6, 6JS6A, 6JS6C	6KD6, 6LB6, 6LF6, 6LX6, 6LR6
6JS6C	12FY	6JS6, 6JS6A, 6JS6B	6KD6, 6LB6, 6LF6, 6LR6, 6LX6
6JT6	9QU	6JT6A	6KV6, 6KV6A, 6JG6, 6JG6A
6JT6A	9QU	6JT6	6KV6, 6KV6A, 6JG6, 6JG6A
6JT8	9DX	6LY8	6EB8, 6GN8, 6JE8, 6KV8
6JU6	9QL	6JF6, 6KM6	6JE6, 6JE6A, 6JE6B, 6JE6C, 6LQ6, 6MJ6

Original	Base Diagram	Preferred Substitution	Good Substitution
6JU8	9PQ	6JU8A	
6JU8A	9PQ	6JU8	
6JV8	9DX	6JE8, 6KS8	6AW8, 6AW8A, 6EB8, 6KR8, 6KV8, 6LY8
6JW8	9AE	6JW8/ECF802, 6LX8	6KE8, 6MG8
6JZ6	12GD		
6JZ8	12DZ		6LU8
6K5	5U	6K5GT	
6K5GT	5U	6K5	
6K6	7S	6K6G, 6K6GT, 6K6MG	6F6, 6F6G, 6F6GT, 6F6MG
6K6G	7S	6K6, 6K6GT, 6K6MG	6F6, 6F6G, 6E6GT, 6F6MG
6K6GT	7S	6K6, 6K6G, 6K6MG	6F6, 6F6G, 6F6GT, 6F6MG
6K6MG	7S	6K6, 6K6G, 6K6GT	6F6, 6F6G, 6F6GT, 6F6MG
6K7	7R	6K7G, 6K7GT, 6K7GTX, 6K7MG, 6S7, 6S7G	6J7, 6J7G, 6J7GT, 6J7GTX, 6J7MG, 6U7, 6U7G
6K7G	7R	6K7, 6K7GT, 6K7GTX, 6K7MG, 6S7, 6S7G	6J7, 6J7G, 6J7GT, 6J7GTX, 6J7MG, 6U7, 6U7G
6K7GT	7R	6K7, 6K7G,	6J7, 6J7G,

Original	Base Diagram	Preferred Substitution	Good Substitution
		6K7GTX, 6K7MG, 6S7, 6S7G	6J7GT, 6J7GTX, 6J7MG, 6U7, 6U7G
6K7GTX	7R	6K7, 6K7G, 6K7GT, 6K7MG, 6S7, 6S7G	6J7, 6J7G, 6J7GT, 6J7GTX, 6J7MG, 6U7, 6U7G
6K7MG	7R	6K7, 6K7G, 6K7GT, 6K7GTX, 6S7, 6S7G	6J7, 6J7G, 6J7GT, 6J7GTX, 6J7MG, 6U7, 6U7G
6K8	8K	6K8G, 6K8GT, 6K8GTX	
6K8G	8K	6K8, 6K8GT, 6K8GTX	
6K8GT	8K	6K8, 6K8G, 6K8GTX	
6K8GTX	8K	6K8, 6K8G, 6K8GT	
6K11	12BY	6Q11	6AV11
6KA8	9PV		
6KD8	9AE	6U8, 6U8A, 6EA8, 6GH8, 6GH8A	6AX8, 6GJ8, 6LM8
6KE8	9DC	6MG8	
6KF8	9FG	6BU8, 6HS8, 6GS8, 6MK8, 6MK8A	
6KG6	9RJ	6KG6A, 6KG6A/EL509	
6KG6A	9RJ	6KG6, 6KG6A/EL509	
6KL8	9LQ		6EQ7
6KM6	9QL	6JF6	6JU6
6KM8	9QG		
6KN6	12GU		
6KN8	9AJ	6KN8/6RHH8, 6FW8	6BK7, 6BK7A,

(continued)

Original	Base Diagram	Preferred Substitution	Good Substitution
			6BK7B, 6BS8, 6BZ7, 6BZ8, 6JK8
6KR8	9DX	6LB8, 6LQ8	6CX8, 6JL8, 6GN8, 6HF8, 6JT8, 6LY8
6KS6	7DF	6BN6	
6KS8	9DX	6AW8, 6AW8A, 6JV8, 6LF8, 6HF8	6BA8, 6BA8A, 6GN8, 6JT8
6KT6	9PM	6JC6, 6JC6A, 6JD6	6HM6, 6HT6
6KT8	9QP		
6KU8	9LT		6ET7
6KV6	9QU	6KV6A, 6JG6, 6JG6A	
6KV6A	9QU	6KV6, 6JG6, 6JG6A	
6KV8	9DX	6JT8, 6KR8, 6LB8, 6LQ8	6EB8, 6JV8, 6LY8
6KY6	9GK		
6KY8	9QT	6KY8A	6LR8
6KY8A	9QT	6KY8	6LR8
6KZ8	9FZ	6CS8	
6L5	6Q	6L5G, 6C5, 6C5G, 6C5GT	6AE5, 6J5, 6J5G, 6J5GT, 6J5GTX, 6J5GX, 6J5MG
6L5G	6Q	6L5, 6C5, 6C5G, 6C5GT	6AE5, 6J5, 6J5G, 6J5GT, 6J5GTX, 6J5GX, 6J5GX, 6J5MG

Original	Base Diagram	Preferred Substitution	Good Substitution
6L6	7AC	6L6A, 6L6G, 6L6GA, 6L6GB, 6L6GC, 6L6GT, 6L6GX, 6L6Y, 6L6W, 6L6WA, 6L6WB, 6L6WGT, 5881	7581
6L6A	7AC	6L6, 6L6G, 6L6GA, 6L6GB, 6L6GC, 6L6GT, 6L6GX, 6L6Y, 6L6W, 6L6WA, 6L6WB, 6L6WGT, 5881	7581
6L6G	7AC	6L6, 6L6A, 6L6GA, 6L6GB, 6L6GC, 6L6GT, 6L6GX, 6L6Y, 6L6W, 6L6WA, 6L6WB, 6L6WGT, 5881	7581
6L6GA	7AC	6L6, 6L6A, 6L6G, 6L6GB, 6L6GC, 6L6GT, 6L6GX, 6L6Y, 6L6W, 6L6WA, 6L6WB, 6L6WGT, 5881	7581
6L6GB	7AC	6L6, 6L6A, 6L6G, 6L6GA, 6L6GC, 6L6GT, 6L6GX, 6L6Y, 6L6W, 6L6WA, 6L6WB, 6L6WGT, 5881	7581
6L6GC	7AC	6L6, 6L6A, 6L6G, 6L6GA, 6L6GB, 6L6GT, 6L6GX, 6L6Y, 6L6W,	7581

(continued)

Original	Base Diagram	Preferred Substitution	Good Substitution
		6L6WA, 6L6WB, 6L6WGT, 5881	
6L6GT	7AC	6L6, 6L6A, 6L6G, 6L6GA, 6L6GB, 6L6GC, 6L6GX, 6L6Y, 6L6W, 6L6WA, 6L6WB, 6L6WGT, 5881	7581
6L6GX	7AC	6L6, 6L6A, 6L6G, 6L6GA, 6L6GB, 6L6GC, 6L6GT, 6L6Y, 6L6W, 6L6WA, 6L6WB, 6L6WGT, 5881	7581
6L6Y	7AC	6L6, 6L6A, 6L6G, 6L6GA, 6L6GB, 6L6GC, 6L6GX, 6L6GT, 6L6W, 6L6WA, 6L6WB, 6L6WGT, 5881	7581
6L6W	7AC	6L6, 6L6A, 6L6G, 6L6GA, 6L6GB, 6L6GC, 6L6GT, 6L6GX, 6L6Y, 6L6WA, 6L6WB, 6L6WGT, 5881	7581
6L6WA	7AC	6L6, 6L6A, 6L6G, 6L6GA, 6L6GB, 6L6GC, 6L6GT, 6L6GX, 6L6Y, 6L6W, 6L6WB, 6L6WGT, 5881	7581
6L6WB	7AC	6L6, 6L6A, 6L6G, 6L6GA, 6L6GB, 6L6GC, 6L6GT, 6L6GX, 6L6Y, 6L6W, 6L6WA, 6L6WGT, 5881	7581

78

Original	Base Diagram	Preferred Substitution	Good Substitution
6L6WGT	7AC	6L6, 6L6A, 6L6G, 6L6GA, 6L6GB, 6L6GC, 6L6GT, 6L6GX, 6L6Y, 6L6W, 6L6WA, 6L6WB, 5881	7581
6L7	8B	6L7G, 1612	
6L7G	8B	6L7, 1612	
6L10	8Y	6AG7, 6AG7Y	
6L12	9AJ	6AQ8, 6AQ8/ECC85	
6L13	9A	12AX7, 12AX7A, 12AX7A/ECC83	
6L31	7BZ	6AQ5, 6AQ5A	
6L43	9BV	6CL6	
6LB6	12JF		6JS6, 6JS6A, 6JS6B, 6JS6C
6LB8	9DX	6JT8	6EB8, 6GN8, 6HF8, 6JE8, 6LY8
6LC6	8ML	6LH6, 6LH6A, 6LJ6, 6LJ6A	
6LC8	9QY		
6LD12	9E	6AK8, 6AK8/EABC80	
6LE8	9QZ		
6LF6	12GW	6LX6, 6MH6	
6LF8	9DX	6JV8, 6KS8	6AW8, 6AW8A
6LH6	8MG	6LH6A, 6LJ6, 6LJ6A	6LC6
6LH6A	8MG	6LH6, 6LJ6, 6LJ6A	6LC6
6LJ6	8MG	6LJ6A, 6LH6, 6LH6A	6LC6

Original	Base Diagram	Preferred Substitution	Good Substitution
6LJ6A	8MG	6LH6, 6LH6A, 6LJ6	6LC6
6LJ8	9GF	6GD7	6MB8
6LM8	9AE	6GH8, 6GH8A	6AX8, 6EA8, 6GJ8, 6HL8, 6KD8, 6U8, 6U8A
6LN8	9DC	6LN8/LCF80, 6BL8, 6BL8/ECF80	6KE8
6LP12	9EX	6BM8/ECL82 6BM8	
6LQ6	9QL	6JE6, 6JE6A, 6JE6B, 6JE6C, 6MQ6	
6LQ8	9DX	6KR8	6CX8, 6JT8, 6KV8
6LR6	12FY	6JS6, 6JS6A, 6JS6B, 6JS6C	
6LR8	9QT		6KY8
6LT8	9RL		
6LU8	12DZ		
6LX6	12GW	6LF6	
6LX8	9AE	6LX8/LCF802, 6JW8, 6JW8/ECF802	6KE8
6LY8	9DX		6EB8, 6GN8, 6HF8, 6JE8, 6JT8, 6KV8, 6KR8
6LZ6	9QL	6ME6	
6M1	6R	6U5	
6M7G			6K7, 6K7G, 6K7GT,

Original	Base Diagram	Preferred Substitution	Good Substitution
			6K7GTX, 6K7MG
6M7G	7R	6M7	6K7, 6K7G, 6K7GT, 6K7GTX, 6K7MG
6M11	12CA		
6MA6	8NP		
6MB8	9FA	6LJ8	6GD7
6MD8	9RQ		
6ME6	9QL	6LQ6, 6JE6, 6JE6A, 6JE6B, 6JE6C, 6MJ6	6LZ6
6ME8	9RU		
6MF8	12DZ		6LU8
6MG8	9DC	6JW8, 6JW8/ECF802, 6LX8, 6LX8/LCF802	6KE8
6MH1		6J4, 6J4A, 6J4WA, 6J4WB	
6MH6	12GW	6LF6, 6LX6	
6MHH3	7BF	6J6, 6J6A, 6J6WA, 6J6WB	
6MJ6	9QL	6LQ6, 6JE6, 6JE6A, 6JE6B, 6JE6C	
6MJ8	12HG		
6MK8	9FG	6MK8A	6BU8, 6KF8
6MK8A	9FG	6MK8	6BU8, 6KF8
6ML8	9RQ		6MD8
6MN8	12HU		
6MQ8	9AE		6HL8
6MU8	9AE		6EA8, 6U8, 6U8A
6MV8	9DX		
6N5	6R	6AB5	6E5, 6U5

Original	Base Diagram	Preferred Substitution	Good Substitution
6N6	7AU	6N6G, 6AB6	
6N6G	7AU	6N6, 6AB6	
6N7	8B	6N7G, 6N7GT, 6N7MG	6Y7, 6Y7G
6N7G	8B	6N7, 6N7GT, 6N7MG	6Y7, 6Y7G
6N7GT	8B	6N7, 6N7G, 6N7MG	6Y7, 6Y7G
6N7MG	8B	6N7, 6N7G, 6N7GT	6Y7, 6Y7G
6N8	9HE		6AD8, 6DC8
6P5	6Q	6P5GT, 6C5, 6C5G, 6C5GT	6AE5, 6AE5GT, 6J5, 6J5G, 6J5GT, 6J5GTX, 6J5GX, 6J5MG
6P5GT	6Q	6P5, 6C5, 6C5G, 6C5GT	6AE5, 6AE5GT 6J5, 6J5G, 6J5GT, 6J5GTX, 6J6GX, 6J5MG
6P7	7U	6P7G	
6P9	7BZ	6AQ5, 6AQ5A, 6HG5	
6P15	9CV	6BQ5, 6BQ5/EL84	
6PL12	9EX	6BM8, 6BM8/ECL82	
6Q7	7V	6Q7G, 6Q7GT, 6Q7MG	6B6, 6B6G, 6T7, 6T7G
6Q7G	7V	6Q7, 6Q7GT, 6Q7MG	6B6, 6B6G, 6T7, 6T7G
6Q7GT	7V	6Q7, 6Q7G,	6B6, 6B6G,
6Q7MG	7V	6Q7, 6Q7G, 6Q7GT	6B6, 6B6G, 6T7, 6T7G

Original	Base Diagram	Preferred Substitution	Good Substitution
6Q8	8A	6A8, 6A8G, 6A8GT	
6Q11	12BY	6K11	6D10
6R3	9CB	6AF3, 6AL3, 6AL3/EY88, 6BR3, 6RK19	
6R7	7V	6R7G, 6R7GT	6B6, 6B6G, 6Q7, 6Q7G, 6Q7GT, 6Q7MG, 6T7, 6T7G
6R7G	7V	6R7, 6R7GT	6B6, 6B6G, 6Q7, 6Q7G, 6Q7GT, 6Q7MG, 6T7, 6T7G
6R7GT	7V	6R7, 6R7G	6B6, 6B6G, 6Q7, 6Q7G, 6Q7GT, 6Q7MG, 6T7, 6T7G
6R8	9E	6T8, 6T8A	
6RHH2	9AJ	6BC8, 6BZ8	
6RHH8	9AJ	6KN8	
6RK19	9CB	6BR3	
6RP22	9BV		
6RR8	9X	6RR8C, 5847/404A	
6RR8C	9X	6RR8, 5847/404A	
6S4	9AC	6S4A	
6S4A	9AC	6S4	
6S5G	6R	6E5	
6S7	7R	6S7G, 6K7, 6K7G, 6K7GT, 6K7GTX, 6K7MG	6U7, 6U7G, 6W7, 6W7G
6S7G	7R	6S7, 6K7, 6K7G, 6K7GT, 6K7GTX, 6K7MG	6U7, 6U7G, 6W7, 6W7G
6S8	8CB	6S8GT	

Original	Base Diagram	Preferred Substitution	Good Substitution
6S8GT	8CB	6S8	
6SA7	8R	6SA7G, 6SA7GT, 6SA7GTX, 6SA7GTX, 6SA7GTY, 6SA7Y, 6SB7, 6SB7GTY, 6SB7Y	
6SA7G	8R	6SA7, 6SA7GT, 6SA7GTX, 6SA7GTY, 6SA7Y, 6SB7, 6SB7GTY, 6SB7Y	
6SA7GT	8AD	6SA7, 6SA7G, 6SA7GTX, 6SA7GTY, 6SA7Y, 6SB7, 6SB7GTY, 6SB7Y	
6SA7GTX	8R	6SA7, 6SA7G, 6SA7GT, 6SA7GTY, 6SA7Y, 6SB7, 6SB7GTY,	
6SA7GTY	8R	6SA7, 6SA7G, 6SA7GT, 6SA7GTX, 6SA7Y, 6SB7, 6SB7GTY, 6SB7Y	
6SA7Y	8R	6SA7, 6SA7G, 6SA7GT, 6SA7GTX, 6SA7GTY, 6SB7, 6SB7GTY, 6SB7Y	
6SB7	8R	6SB7GTY, 6SB7Y	
6SB7GTY	8R	6SB7, 6SB7Y	
6SB7Y	8R	6SB7, 6SB7GTY	
6SC7	8S	6SC7GT, 6SC7GTY	
6SC7GT	8S	6SC7, 6SC7GTY	
6SC7GTY	8S	6SC7, 6SC7GT	

Original	Base Diagram	Preferred Substitution	Good Substitution
6SD7	8N	6SK7, 6SK7G, 6SK7GT, 6SK7GTY, 6SK7Y, 6SK7W, 6SK7WA, 6SK7WGT	6AB7, 6AC7, 6AC7W, 6AJ7
6SE7	8N	6SK7, 6SK7G, 6SK7GT, 6SK7GTX, 6SK7GTY, 6SK7Y, 6SK7W, 6SK7WA, 6SK7WGT	6AB7, 6AC7, 6AC7W, 6AJ7
6SF5	6AB	6SF5GT	
6SF5GT	6AB	6SF5	
6SF7	7AZ		6SV7
6SG7	8BK	6SG7GT, 6SG7Y	6SH7, 6SH7GT, 6SH7L, 6AB7, 6AC7, 6AC7W, 6AJ7
6SG7GT	8BK	6SG7, 6SG7Y	6SH7, 6SH7GT, 6SH7L, 6AB7, 6AC7, 6AC7W, 6AJ7
6SG7Y	8BK	6SG7, 6SG7GT	6SH7, 6SH7GT, 6SH7L, 6AB7, 6AC7, 6AC7W, 6AJ7
6SH7	8BK	6SH7GT, 6SH7L	6SG7, 6SG7GT, 6SG7Y,

(continued)

Original	Base Diagram	Preferred Substitution	Good Substitution
			6AB7, 6AC7, 6AC7W, 6AJ7
6SH7GT	8BK	6SH7, 6SH7L	6SG7, 6SG7GT, 6SG7Y, 6AB7, 6AC7, 6AC7W, 6AJ7
6SH7L	8BK	6SH7, 6SH7GT	6SG7, 6SG7GT, 6SG7Y, 6AB7, 6AC7, 6AC7W, 6AJ7
6SJ7	8N	6SJ7GT, 6SJ7GTX, 6SJ7GTY, 6SJ7Y, 5963	6AB7, 6AC7, 6AC7W, 6AJ7, 6SK7, 6SK7G, 6SK7GT, 6SK7GTX, 6SK7GTY, 6SK7Y, 6SK7W, 6SK7WA, 6SK7WGT
6SJ7GT	8N	6SJ7, 6SJ7GTX, 6SJ7GTY, 6SJ7Y, 5963	6AB7, 6AC7, 6AC7W, 6AJ7, 6SK7, 6SK7G, 6SK7GT, 6SK7GT, 6SK7GTX, 6SK7GTY,

Original	Base Diagram	Preferred Substitution	Good Substitution
			6SK7Y, 6SK7W, 6SK7WA, 6SK7WGT
6SJ7GTX	8N	6SJ7, 6SJ7GT, 6SJ7GTY, 6SJ7Y, 5963	6AB7, 6AC7, 6AC7W, 6AJ7, 6SK7, 6SK7G, 6SK7GT, 6SK7GTX, 6SK7GTY, 6SK7Y, 6SK7W, 6SK7WA, 6SK7WGT
6SJ7GTY	8N	6SJ7, 6SJ7GT, 6SJ7GTX, 6SJ7Y, 5963	6AB7, 6AC7, 6AC7W, 6AJ7, 6SK7, 6SK7G, 6SK7GT, 6SK7GTX, 6SK7GTY 6SK7Y, 6SK7W, 6SK7WA, 6SK7WGT
6SJ7GTY	8N	6SJ7, 6SJ7GT, 6SJ7GTX, 6SJ7Y, 5963	6AB7, 6AC7, 6AC7W, 6AJ7, 6SK7GT, 6SK7GTX, 6SK7GTY, 6SK7Y, 6SK7W, 6SK7WA, 6SK7WGT
6SY7Y	8N	6SJ7GT,	6AB7, 6AC7,

(continued)

Original	Base Diagram	Preferred Substitution	Good Substitution
		6SJ7GTX, 6SJ7GTY, 5963	6AC7W, 6SK7, 6SK7G, 6SK7GT, 6SK7GTX, 6SK7GTY, 6SK7Y, 6SK7W, 6SK7WA, 6SK7WGT
6SK7	8N	6SK7G, 6SK7GT, 6SK7GTX, 6SK7GTY, 6SK7Y, 6SK7W, 6SK7WA 6SK7WGT	6AB7, 6AC7, 6AC7W, 6AJ7, 6SS7, 6SS7GT
6SK7G	8N	6SK7, 6SK7GT, 6SK7GTX, 6SK7GTY, 6SK7Y, 6SK7W, 6SK7WA, 6SK7WGT	6AB7, 6AC7, 6AC7W, 6AJ7, 6SS7, 6SS7GT
6SK7GT	8N	6SK7, 6SK7G, 6SK7GTX, 6SK7GTY, 6SK7Y, 6SK7W, 6SK7WA, 6SK7WGT	6AB7, 6AC7, 6AC7W, 6AJ7, 6SS7, 6SS7GT
6SK7GTX	8N	6SK7, 6SK7G, 6SK7GT, 6SK7GTY, 6SK7Y, 6SK7W, 6SK7WA, 6SK7WGT	6AB7, 6AC7, 6AC7W, 6AJ7, 6SS7, 6SS7GT
6SK7GTY	8N	6SK7, 6SK7G, 6SK7GT, 6SK7GTX, 6SK7Y, 6SK7W,	6AB7, 6AC7, 6AC7W, 6AJ7, 6SS7,

Original	Base Diagram	Preferred Substitution	Good Substitution
		6SK7WA, 6SK7WGT	6SS7GT
6SK7Y	8N	6SK7, 6SK7G, 6SK7GT, 6SK7GTX, 6SK7GTY, 6SK7W, 6SK7WA, 6SK7WGT	6AB7, 6AC7, 6AC7W, 6AJ7, 6SS7, 6SS7GT
6SK7W	8N	6SK7, 6SK7G, 6SK7GT, 6SK7GTX, 6SK7GTY, 6SK7Y 6SK7WA, 6SK7WGT	6AB7, 6AC7, 6AC7W, 6AJ7 6SS7, 6SS7GT
6SK7WA	8N	6SK7, 6SK7G, 6SK7GT, 6SK7GTX, 6SK7GTY, 6SK7Y, 6SK7W, 6SK7WGT	6AB7, 6AC7, 6AC7W, 6AJ7, 6SS7, 6SS7GT
6SK7WGT	8N	6SK7, 6SK7G, 6SK7GT, 6SK7GTX, 6SK7GTY, 6SK7Y 6SK7W, 6SK7WA	6AB7, 6AC7, 6AC7W, 6AJ7, 6SS7, 6SS7GT
6SL7	8BD	6SL7A, 6SL7GT, 6SL7GTY, 6SL7L, 5691	6SN7, 6SN7A, 6SN7GTA, 6SN7GTB, 6SN7GTY, 6SN7L
6SL7A	8BD	6SL7, 6SL7GT, 6SL7GTY, 6SL7L, 5691,	6SN7, 6SN7A, 6SN7GTA, 6SN7GTB, 6SN7GTY, 6SN7L
6SL7GT	8BD	6SL7, 6SL7A, 6SL7GTY, 6SL7L, 5691	6SN7, 6SN7A, 6SN7GTA, 6SN7GTB, 6SN7GTY, 6SN7L

Original	Base Diagram	Preferred Substitution	Good Substitution
6SL7GTY	8BD	6SL7, 6SL7A, 6SL7GT, 6SL7L, 5691	6SN7, 6SN7A, 6SN7GTA, 6SN7GTB, 6SN7GTY, 6SN7L
6SL7L	8BD	6SL7, 6SL7A, 6SL7GT, 6SL7GTY, 6591	6SN7, 6SN7A, 6SN7GTA, 6SN7GTB, 6SN7GTY, 6SN7L
6SN7	8BD	6SN7A, 6SN7GTA, 6SN7GTB, 6SN7GTY, 6SN7L 5692	6BL7, 6BL7GT, 6BL7GTA, 6BX7, 6BX7GT, 6DN7, 6EA7, 6EM7, 6GL7
6SN7A	8BD	6SN7, 6SN7GTA, 6SN7GTB 6SN7GTY, 6SN7L, 5692	6BL7, 6BL7GT, 6BL7GTA, 6EA7, 6BX7, 6BX7GT, 6DN7, 6EM7, 6GL7
6SN7GTA	8BD	6SN7, 6SN7A, 6SN7GTB, 6SN7GTY, 6SN7L, 5692	6BL7, 6BL7GT, 6BL7GTA, 6BX7, 66BX6GT, 6DN7, 6EA7, 6EM7, 6GL7
6SN7GTB	8BD	6SN7, 6SN7A, 6SN7GTA,	6BL, 6BL7GT, 6BL7GTA, 6BX7,

Original	Base Diagram	Preferred Substitution	Good Substitution
		6SN7GTY, 6SN7L, 5692	6BX7GT, 6DN7, 6EA7, 6EM7, 6GL7
6SN7GTY	8BD	6SN7, 6SN7A, 6SN7GTA, 6SN7GTB, 6SN7L, 5692	6BL7, 6BL7GT, 6BL7GTA, 6BX7, 6BX7GT, 6DN7, 6EA7, 6EM7, 6GL7
6SN7L	8BD	6SN7, 6SN7A, 6SN7GTA, 6SN7GTB,	6BL7, 6BL7GT, 6BL7GTA, 6BX7, 6BX7GT, 6DN7, 5692 6EA7, 6EM7, 6HL7
6SQ7	8Q	6SQ7G, 6SQ7GT	6SZ7
6SQ7G	8Q	6SQ7, 6SQ7GT	6SZ7
6SQ7GT	8Q	6SQ7, 6SQ7G	6SZ7
6SR7	8Q	6SR7G, 6SR7GT, 6ST7	6SZ7
6SR7G	8Q	6SR7, 6SR7GT, 6ST7	6SZ7
6SR7GT	8Q	6SR7, 6SR7G, 6ST7	6SZ7
6SS7	8N	6SS7GT, 6SK7, 6SK7G, 6SK7GT, 6SK7GTX, 6SK7GTY, 6SK7Y, 6SK7W, 6SK7WA, 6SK7WGT	6SD7
6SS7GT	8N	6SS7, 6SK7, 6SK7G, 6SK7GT, 6SK7GTX, 6SK7GTY, 6SK7Y, 6SK7W, 6SK7WA, 6SK7WGT	6SD7
6ST7	8Q	6SR7, 6SR7G, 6SR7GT	6SQ7, 6SQ7G, 6SQ7GT, 6SZ7
6SU7	8BD	6SL7, 6SL7A, 6SL7GT, 6SL7GTY, 6SL	6SN7, 6SN7A, 6SN7GTA, 6SN7GTY, 6SN7L

Original	Base Diagram	Preferred Substitution	Good Substitution
6SV7	7AZ		6SF7
6SZ7	8Q	6SQ7, 6SQ7G, 6SQ7GT	
6T4	7DK	6AF4, 6AF4A, 6DZ4	
6T5	6R		6E5, 6U5, 6AB5, 6N5
6T7	7V	6T7G, 6Q7, 6Q7G, 6Q7GT, 6Q7MG	6B6, 6B6G
6T7G	7V	6T7, 6Q7, 6Q7G, 6Q7GT, 6Q7MG	6B6, 6B6G
6T8	9E	6T8A	6AK8, 6AK8/EABC80, 6R8
6T8A	9E	6T8	6AK8, 6AK8/EABC80 6R8
6T9	12FM		
6T10	12EZ	6AD10	
6T35	750TL		
6U4	4CG	6U4GT, 6AS4, 6AX4, 6AX4GT, 6AX4GTB, 6DA4, 6DM4, 6DM4A, 6DQ4, 6DT4, 6DT4, 6W4, 6W4GT	
6U4GT	4CG	6U4, 6AS4, 6AX4, 6AX4GT, 6AX4GTB, 6DA4, 6DM4, 6DM4A, 6DQ4, 6DT4, 6W4, 6W4GT	
6U5	6R	6E5	6T5, 6AB5, 6N5
6U6	7AC	6U6GT, 6Y6, 6Y6G, 6Y6GA, 6Y6GT	6EY6, 6EZ5
6U6GT	7AC	6U6, 6Y6, 6Y6G, 6Y6GA, 6Y6GT	6EY6, 6EZ5

92

Original	Base Diagram	Preferred Substitution	Good Substitution
6U7	7R	6U7G, 6K7, 6K7G, 6K7GT, 6K7GTX, 6K7MG	6S7, 6S7G
6U7G	7R	6U7, 6K7, 6K7G, 6K7GT, 6K7GTX, 6K7MG	6S7, 6S7G
6U8	9AE	6U8A, 6GH8, 6GH8A, 6EA8, 6KD8	6GJ8, 6LM8
6U8A	9AE	6U8, 6GH8, 6GH8A, 6EA8, 6KD8	6GJ8, 6LM8
6U9	10K	6U9/ECF201	6X9, 6X9/ECF200
6U10	12FE	6AC10	
6V3	9BD	6V3A	
6V3A	9BD	6V3	
6V4	9M		6CA4
6V6	7AC	6V6G, 6V6GT, 6V6GTA, 6V6GTX, 6V6GX, 6V6Y, 6V6GTY, 7408	1614
6V6G	7AC	6V6, 6V6GT, 6L6GB, 6V6GTA, 6V6GTX, 6V6GTY, 6V6GX, 6V6Y, 7408	1614
6V6GT	7AC	6V6, 6V6G, 6V6GTA, 6V6GTX, 6V6GTY, 6V6GX, 6V6Y, 7408	1614
6V6GTA	7AC	6V6, 6V6G, 6V6GT, 6V6GTX, 6V6GTY, 6V6GX, 6V6Y, 7408	1614
6V6GTX	7AC	6V6, 6V6G, 6V6GT, 6V6GTA, 6V6GTY, 6V6GX, 6V6Y, 7408	1614

Original	Base Diagram	Preferred Substitution	Good Substitution
6V6GTY	7AC	6V6, 6V6G, 6V6GT, 6V6GTA, 6V6GTX, 6V6GX, 6V6Y, 7408	1614
6V6GX	7AC	6V6, 6V6G, 6V6GT, 6V6GTA, 6V6GTX, 6V6GTY, 6V6Y, 7408	1614
6V6Y	7AC	6V6, 6V6G, 6V6GT, 6V6GTA, 6V6GTX, 6V6GTY, 6V6GX, 7408	1614
6V7	7R	6V7G	6R7, 6R7G, 6R7GT
6V7G	7R	6V7	6R7, 6R7G, 6R7GT
6W4	4CG	6W4GT, 6AS4, 6AX4, 6AX4GT, 6AX4GTB, 6DA4, 6DM4, 6DM4A, 6DQ4, 6DT4	6U4, 6U4GT
6W4GT	4CG	6W4, 6AS4, 6AX4, 6AX4GT, 6AX4GTB, 6DA4, 6DM4, 6DM4A, 6DQ4, 6DT4	6U4, 6U4GT
6W5	6S	6W5G, 6W5GT, 6AX5, 6AX5GT	
6W5G	6S	6W5, 6W5GT, 6AX5, 6AX5GT	
6W5GT	6S	6W5, 6W5G, 6AX5, 6AX5GT	
6W6	7AC	6W6GT	6EY6, 6EZ5, 6DG6, 6F6, 6F6G, 6F6GT, 6F6MG
6W6GT	7AC	6W6	6EY6, 6EZ5, 6DG6, 6F6, 6F6G, 6F6GT, 6F6MG
6W7	7R	6W7G, 6J7GT, 6J7GTX, 6J7MG	6K7, 6K7G, 6K7GT, 6K7GTX, 6K7MG

Original	Base Diagram	Preferred Substitution	Good Substitution
6X4	5BS	6BX4, 6X4W	6202, 6AV4
6X4W	5BS	6X4, 6BX4	6202, 6AV4
6X5	6S	6X5G, 6X5GT, 6X5L, 6X5MG, 6X5W, 6X5WGT, 6W5, 6W5G, 6W5GT	6AX5, 6AX5GT
6X5G	6S	6X5, 6X5GT, 6X5L, 6X5MG, 6X5W, 6X5WGT, 6W5, 6W5G, 6W5GT	6AX5, 6AX5GT
6X5GT	6S	6X5, 6X5G, 6X5L, 6X5MG, 6X5W, 6X5WGT, 6W5, 6W5G, 6W5GT	6AX5, 6AX5GT
6X5L	6S	6X5, 6X5G, 6X5GT, 6X5MG, 6X5W, 6X5WGT, 6W5, 6W5G, 6W5GT	6AX5, 6AX5GT
6X5MG	6S	6X5, 6X5G, 6X5GT, 6X5L, 6X5W, 6X5WGT 6W5, 6W5G, 6W5GT	6AX5, 6AX5GT
6X5W	6S	6X5, 6X5G, 6X5GT, 6X5L, 6X5MG, 6X5WGT, 6W5, 6W5G, 6W5GT	6AX5, 6AX5GT
	6S	6X5, 6X5G, 6X5GT, 6X5L, 6X5MG, 6X5W, 6W5, 6W5G, 6W5GT	6X5WGT 6AX5, 6AX5GT
6X8	9AK	6X8A	

Original	Base Diagram	Preferred Substitution	Good Substitution
6X8A	9AK	6X8	
6X9	10K	6X9/ECF200	6U9, 6U9 ECF201
6Y5	6J		
6Y6	7AC	6Y6G, 6Y6GA, 6Y6GT, 6U6, 6U6GT	6EZ5
6Y6G	7AC	6Y6, 6Y6GA, 6Y6GT, 6U6, 6U6GT	6EZ5
6Y6GA	7AC	6Y6, 6Y6G, 6Y6GT, 6U6, 6U6GT	6EZ5
6Y6GT	7AC	6Y6, 6Y6G, 6Y6GA, 6U6, 6U6GT	6EZ5
6Y7	8B	6Y7G	6N7, 6N7G, 6N7GT, 6N7MG
6Y7G	8B	6Y7	6N7, 6N7G, 6N7GT, 6N7MG
6Y9	10L	6Y9/EFL200	
6Y10	12EZ		6AD10, 6T10
6Z3	4G	1V	
6Z4	5D		84,98
6Z5	6K		
6Z7G	8B		
6Z10	12BT	6J10	
6Z31	5BS	6X4, 6X4W	6202
6ZY5	6S	6ZY5G, 6X5, 6X5G, 6X5GT, 6X5L, 6X5MG, 6X5W, 6X5WGT	
6ZY5G	6S	6ZY5, 6X5, 6X5G, 6X5GT, 6X5L, 6X5MG, 6X5W, 6X5WGT	
7A4	5AC		

96

Original	Base Diagram	Preferred Substitution	Good Substitution
7A5	6AA		
7A6	7AJ		
7A7	8V	7B7, 7L7	7AH7, 7G7, 7H7, 7T7
7A8	8U		7B8
7AD7	8V		
7AF7	8AC	7N7	
7AG7	8V	7G7	7AH7, 7H7, 7T7, 7V7
7AH7	8V	7AG7	7G7, 7H7, 7T7, 7V7
7AJ7	8V	7C7, 7L7	7A7, 7B7, 7G7
7AK7	8V		7G7, 7T7
7AU7	9A		6AU7, 6AX7
7B4	5AC		7A4
7B5	6AE		7C5
7B6	8W	7C6	
7B7	8V	7AH7	7A7, 7AJ7, 7H7, 7T7
7B8	8X		7A8
7C5	6AA		7B5
7C6	8W		7B6
7C7	8V	7B7	7A7, 7AH7, 7AJ7, 7G7, 7H7, 7L7
7D11	8ET	6CA7, 6CA7/EL34, 6550	
7DJ8	9DE	7DJ8/PCC88	6GM8
7E5	8BN		
7E6	8W		
7E7	8AE	7R7	
7EY6	7AC		
7F7	8AC		
7F8	8BW		
7F25		8166, 4-1000A	
7F25A		8166, 4-1000A	

Original	Base Diagram	Preferred Substitution	Good Substitution
7G7	8V	7AG7	6AH7, 7H7, 7L7
7GS7	9GF		6GS7
7H7	8V	7T7	7A7, 7AG7, 7AH7, 7V7
7HG8	9MP	7HG8/PCF88	8HG8
7J7	8BL	7S7	
7K7	8BF		
7KY6	9GK	7KZ6	
7KZ6	9GK	7KY6	
7L7	8V	7G7, 7H7	7A7, 7AH7, 7T7, 7V7
7N7	8AC	7AF7	
7Q7	8AL		
7R7	8AE		7E7
7S7	8BL		7J7
7T7	8V	7AG7, 7H7, 7V7	7A7, 7AH7
7T40		1000T	
7T45		1500T	
7V7	8V	7T7	7AG7
7W7	8BJ		
7X7	8BZ		
7Y4	5AB		7Z4
7Z4	5AB		
8A8	9DC	9A8, 9A8/PCF80	
8AC10	12FE		
8AL9	12HE		
8AR11	12DM	8BQ11	
8AU8	9DX	8AW8, 8AW8A, 8BA8, 8BA8A	8BH8, 8JT8, 8KS8
8AW8	9DX	8AW8A, 8AU8, 8KS8	8BA8, 8BH8, 8JV8
8AW8A	9DX	8AW8, 8AU8, 8KS8	8BA8, 8JV8, 8BH8
8B8	9EX		
8B10	12BF		

Original	Base Diagram	Preferred Substitution	Good Substitution
8BA8	9DX	8BH8, 8BA8A	8AU8, 8AW8, 8AW8A, 8JV8, 8KS8
8BA8A	9DX	8BA8, 8BH8	8AU8, 8AW8, 8AW8A, 8JV8, 8KS8
8BA11	12ER		
8BH8	9DX	8AU8	8AW8, 8AW8A, 8JV8, 8KS8
8BM11	12FU		
8BN8	9ER		
8BN11	12GF		
8BQ5	9CV		8CW5, 8CW5/XL86
8BQ11	12DM	8AR11	
8BU11	12FP		
8CB11	12DM		
8CG7	9AJ	8FQ7	8GU7
8CM7	9ES		
8CN7	9EN		
8CS7	9EF		
8CW5	9CV	8CW5/XL86	
8CX8	9DX		8EB8, 8GN8, 8JE8
8EB8	9DX	8GN8	8CX8, 8JE8
8EM5	9HN		
8ET7	9LT		
8F10R		8170, 4CX5000A	
8F11R		8171, 4CX10000D	
8FQ7	9LP	8CG7	
8GJ7	9QA	8GX7, 8GJ7/PCF801	
8GN8	9DX	8EB8	8CX8, 8JE8
8GU7	9LP		8FQ7, 8CG7
8GX7	9QA	8GJ7, 8GJ7/PCF801	

Original	Base Diagram	Preferred Substitution	Good Substitution
9GV8	9LY	9GV8/XCL85	
9JW8	9AE	9JW8/PCF802	
9KC6	9RF		
9KX6	9GK	9LA6	
9KZ8	9FZ		
9LA6	9GK	9KX6	
9MN8	12HU		
9RAL1	9HF	12DE7, 12EW7	
9T69		6696A	
8HG8	9MP	7HG8	
8JE8	9DX	8CX8	8EB8, 8GN8, 8JV8
8JU8A	9PQ		
8JV8	9DX	8AW8, 8AW8A	8AU8, 8KS8
8KA8	9PV		
8KS8	9DX	8AW8, 8AW8A	8AU8, 8JV8
8LC8	9QY		
8LT8	9RL		
8U9	10K	8X9	
8X9	10K		
9A8	9DC	9A8/PCF80, 8A8	9EA8, 9U8, 9U8A, 9GH8, 9GH8A
9AH9	12HJ		
9AK10	12FE		9AM10
9AM10	12FE		9AK10
9AQ8	9DE	9AQ8/PCC85	
9AU7	9A		
9BJ11	12FU		
9BR7	9CF		
9CL8	9FX		
9EA8	9AE	9GH8, 9GH8A, 9U8, 9U8A	8A8, 9A8, 9A8/PCF80
9GH8	9AE	9EA8, 9GH8A, 9U8, 9U8A	
9GH8A	9AE	9GH8, 9EA8, 9U8, 9U8A	

Original	Base Diagram	Preferred Substitution	Good Substitution
9U8	9AE	9U8A, 9EA8	
10	4D		
10AL11	12BU		
10BQ5	9CV		10CW5, 10CW5/LL86
10C8	9DA		
10CW5	9CV	10CW5/LL86	
10D2	6BT	12AL5	
10DE7	9HF	10EW7	10DR7, 10FD7, 10FR7
10DR7	9HF	10FD7, 10FR7	
10DX8	9HX	10DX8/LCL84	
10EB8	9DX	10GN8	10HF8, 10JA8, 10JE8, 10LW8
10EG7	8BD	10EM7	
10EM7	8BD	10EG7	
10EW7	9HF	10DE7	10FD7
10FD7	9HF	10FR7	10DR7
10FR7	9HF	10FD7	10DR7
10GF7	9QD	10GF7A	
10GF7A	9QD	10GF7	
10GK6	9GK		
10GN8	9DX	10EB8	10HF8, 10JA8, 10LZ8, 10JY8
10GV8	9LY	10GV8/LCL85	
10HF8	9DX	10EB8	10JA8, 10JY8, 10KR8, 10LW8, 10LZ8
10JA5	12EY		
10JA8	9DX	10LZ8, 10HF8	10EB8, 11KV8
10JT8	9DX	10LW8	10LB8, 10LZ8, 10JA8, 11KV8
10JY8	9DX	10KR8	10EB8, 11LQ8
10KR8	9DX	10LW8	10LB8, 10LZ8, 10JA8, 11KV8
10JY8	9DX	10KR8	10EB8, 11LQ8

Original	Base Diagram	Preferred Substitution	Good Substitution
10KR8	9DX	10JY8, 11LQ8	10LB8, 10LW8
10KU8	9LT		
10LB8	9DX		10KR8, 10HF8, 10JT8, 10JY8
10LE8	9QZ		
10LW8	9DX	10JA8	10EB8, 10JT8, 11KV8
10LY8	9DX		10JT8, 10JY8, 10JA8, 10LZ8
10LZ8	9DX	10JA8	10JT8, 10JY8, 10LY8
10PL12	9EX	50BM8, 50BM8/UCL82	
10T10	12EZ		
10Z10	12BT		
11	4F		
11AF9	10L		
11AR11	12DM	11BQ11	
11BM8	9EX		
11BQ11	12DM	11AR11	
11BT11	12GS		
11C5	7CV		12DM5, 12C5, 12CU5
11CA11	12HN		
11CF11	12HW		
11CH11	12GS		
11CY7	9LG		
11DS5	7BZ		
11FY7	12EQ		
11HM7	9BF		
11JE8	9DX	11KV8, 10LW8, 10EB8	10JY8, 10GN8
11KV8	9DX	11JE8	10LZ8, 10LW8, 10LB8, 10JT8, 10JA8, 10GN8, 10EB8

Original	Base Diagram	Preferred Substitution	Good Substitution
11LQ8	9DX	10KR8	10JY8, 10LB8
11LT8	9RL		
11MS8	9LY		
11Y9	10L	11Y9/LFL200	
12A4	9AG		12B4
12A5	7F		
12A6	7AC	12V6, 12A6Y	
12A6Y	7AC	12A6, 12V6	
12A7	7K		
12A8	8A	12A8GT	
12A8GT	8A	12A8	
12AB5	9EU		7061
12AC6	7BK	12AF6, 12BL6	12CY6
12AC10	12FE		
12AD6	7CH	12AG6, 12FA6	12GA6
12AD7	9A	12AX7, 12AX7A, 12AX7A/ECC83, 7025, 12DF7, 12DM7, 12DT7	12BZ7
12AE6	7BT	12AE6A, 12FT6	12AJ6, 12FM6
12AE6A	7BT	12AE6, 12FT6	12AJ6, 12FM6
12AE7	9A		12U7
12AE10	12EZ	12V10	12BF11
12AF3	9CB	12BR3, 12BY3, 12RK19	
12AF6	7BK	12BL6	
12AG6	7CH	12AD6	12FA6, 12GA6
12AH7GT	8BE		
12AJ6	7BT		12AE6, 12FM6, 12FT6
12AL5	6BT	10D2	
12AL8	9GS		
12AL11	12BU		
12AQ5	7BZ		12BM5
12AS5	7CV	12CA5, 12R5	

Original	Base Diagram	Preferred Substitution	Good Substitution
12AT6	7BT	12BT6	12AV6, 12BK6
12AT7	9A	12AT7/ECC81, 12AT7WA, 12AT7WB, 12AZ7, 12AZ7A	12FV7, 6679, 6201
12AT7WA	9A	12AT7/ECC81, 12AT7, 12AT7WB, 12AZ7, 12AZ7A	12FV7, 6201, 6679
12AT7WB	9A	12AT7, 12AT7/ECC81, 12AT7WA, 12AZ7, 12AZ7A	12FV7, 6201, 6679
12AU6	7BK	12BA6, 12AW6	
12AU7	9A	12AU7A, 12AU7A/ECC82, 12AU7WA, 6189	6670, 5814A, 12AT7, 12AT7WA, 12AT7WB, 12AT7/ECC81, 12AZ7, 12AZ7A, 12BH7
12AU7A	9A	12AU7, 12AU7A/ECC82, 12AU7WA, 6189	6670, 5814A, 12AT7, 12AT7WA, 12AT7WB, 12AT7/ECC81, 12AZ7, 12AZ7A, 12BH7
12AU7WA	9A	12AU7, 12AU7A, 12AU7A/ECC82, 6189	6670, 5814A, 12AT7, 12AT7WA, 12AT7WB, 12AT7/ECC81, 12AZ7, 12AZ7A, 12BH7
12AV5GA	6CK		
12AV6	7BT		12BX6, 12AT6
12AV7	9A		12FV7, 12DW7, 12AZ7, 12AZ7A
12AW6	7CM		12BZ6
12AX3	12BL		12BT3

Original	Base Diagram	Preferred Substitution	Good Substitution
12AX4	4CG	12AX4GT, 12AX4GTA, 12AX4GTB, 12DQ4, 12DM4, 12DM4A	12D4, 12D4A
12AX4GT	4CG	12AX4, 12AX4GTA, 12DQ4, 12DM4, 12DM4A	12D4, 12D4A
12AX4GTA	4CG	12AX4, 12AX4GT, 12AX4GTB, 12DQ4, 12DM4, 12DM4A	12D4, 12D4A
12AX4GTB	4CG	12AX4, 12AX4GT, 12AX4GTA, 12DQ4, 12DM4, 12DM4A	12D4, 12D4A
12AX7	9A	12AX7A, 12AX7A/ECC83, 12AX7WA, 6681, 12DF7, 12DM7, 12DT7, 12DW7	12AD7, 7025
12AX7A	9A	12AX7, 12AX7A/ECC83 12AX7WA, 6681, 12DF7, 12DM7, 12DT7, 12DW7	12AD7, 7025
12AX7WA	9A	12AX7, 12AX7A, 12AX7A/ECC83, 6681, 12DF7, 12DT7, 12DM7, 12DW7	12AD7, 7025
12AY3	9HP	12AY3A, 12BS3, 12BS3A, 12DW4, 12DW4A, 12CK3, 12CL3	
12AY3A	9HP	12AY3, 12BS3, 12BS3A, 12DW4,	

(continued)

Original	Base Diagram	Preferred Substitution	Good Substitution
		12DW4A, 12CK3, 12CL3	
12AY7	9A	12AX7, 12AX7A, 12AX7A/ECC83, 12AX7WA, 12BZ7	12AD7, 12DF7, 12DM7, 12DT7, 12DW7
12AZ7	9A	12AT7, 12AT7WA, 12AT7WB, 12AT7/ECC81, 12AZ7A	12AU7, 12AU7A, 12AU7WA, 12AU7A/ECC82
12AZ7A	9A	12AT7, 12AT7WA, 12AT7WB, 12AT7/ECC81, 12AZ7	12AU7, 12AU7A, 12AU7WA, 12AU7/ECC82
12B4	9AG	12B4A	
12B4A	9AG	12B4	
12B8	8T	12B8GT	
12B8GT	8T	12B8	
12BA6	7BK	12AU6	
12BA7	8CT		
12BB14	9NH	12GB5, 12GB5/XL500	
12BC22	7BT	12AV6	
12BC32	7BT	12AV6	
12BD6	7BK	12BA6	
12BE3	12GA		
12BE6	7CH		12CS6
12BF6	7BT	12BU6	
12BF11	12EZ	12T10	12AE10, 12V10
12BH7	9A	12BH7A, 12DW7	12AU7, 12AU7A, 12AU7WA, 12AU7A/ECC82, 6189
12BH7A	9A	12BH7, 12DW7	12AU7, 12AU7A, 12AU7WA, 12AU7A/ECC82, 6189

Original	Base Diagram	Preferred Substitution	Good Substitution
12BK5	9BQ		
12BK6	7BT	12AV6	12AT6, 12BT6
12BL6	7BK		12AF6, 12CY6, 12EZ6
12BN6	7DF		
12BQ6	6AM	12BQ6GA, 12BQ6GT, 12BQ6GTA, 12BQ6GTB, 12CU6	12DQ6, 12DQ6A, 12DQ6B, 12GW6
12BQ6GA	6AM	12BQ6, 12BQ6GT, 12BQ6GTA, 12BQ6GTB, 12CU6	12DQ6, 12DQ6A, 12DQ6B, 12GW6
12BQ6GT	6AM	12BQ6, 12BQ6GA, 12BQ6GTA, 12BQ6GTB, 12CU6	12DQ6, 12DQ6A, 12DQ6B, 12GW6
12BQ6GTA	6AM	12BQ6, 12BQ6GA, 12BQ6GT, 12BQ6GTB, 12CU6	12DQ6, 12DQ6A, 12DQ6B, 12GW6
12BQ6GTB	6AM	12BQ6, 12BQ6GA, 12BQ6GT, 12BQ6GTA, 12CU6	12DQ6, 12DQ6A, 12DQ6B, 12GW6
12BR3	9CB	12AF3, 12RK19	12BY3
12BR7	9CF		
12BS3	9HP	12BS3A, 12DW4, 12DW4A, 12CL3, 12CK3	12AY3, 12AY3A
12BS3A	9HP	12BS3, 12DW4, 12DW4A, 12CK3, 12CL3	12AY3, 12AY3A
12BT3	12BL	12AX3	
12BT6	7BT	12AT6, 12BK6, 12AV6	
12BU6	7BT	12BF6	
12BV7	9BF	12BY7, 12BY7A, 12DQ7	12GN7, 12GN7A

Original	Base Diagram	Preferred Substitution	Good Substitution
12BV11	12HB		
12BW4	9DJ		
12BY3	9CB	12AF3, 12BR3, 12RK19	
12BY7	9BF	12BY7A, 12BV7, 12DQ7	12GN7, 12GN7A
12BY7A	9BF	12BY7, 12BV7, 12DQ7	12GN7, 12GN7A
12BZ6	7CM	12DX6	
12BZ7	9A		12AX7, 12AX7A, 12AX7WA, 12AX7/ECC83, 6681, 12AY7, 12DW7
12C5	7CV	12CU5	12AS5, 12R5
12C8	8E		
12CA5	7CV		12AS5, 12C5, 12CU5, 12DM5, 12ED5, 12EH5, 12FX5
12CK3	9HP	12CL3, 12DW4, 12DW4A	12BS3, 12BS3A
12CL3	9HP	12DW4, 12DW4A	12CK3
12CM6	9CK		12DW5
12CN5	7CV		
12CR6	7EA		
12CS6	7CH	12BE6	
12CT3	9RX		
12CT8	9DA		
12CU5	7CV	12C5	12DM5, 12R5, 12AS5
12CU6	6AM	12BQ6, 12BQ6GA, 12BQ6GT, 12BQ6GTA, 12BQ6GTB	12DQ6, 12DQ6A, 12DQ6B, 12GW6
12CX6	7BK	12BL6, 12AF6	12DZ6, 12EA6, 12EK6

Original	Base Diagram	Preferred Substitution	Good Substitution
12CY6	7BK	12BL6	12AF6, 12EZ6
12D4	4CG	12D4A, 12DM4, 12DM4A	12AX4, 12AX4GT, 12AX4GTA, 12AX4GTB, 12DQ4
12D4A	4CG	12D4, 12DM4, 12DM4A	12AX4, 12AX4GT, 12AX4GTA, 12AX4GTB, 12DQ4
12DB5	9GR	12CS5	
12DE8	9HG		
12DF7	9A	12AX7, 12AX7A, 12AX7WA, 12AX7A/ECC83, 7025	12AY7, 12BZ7, 12DW7
12DK6	7CM	12BZ6	
12DK7	9HZ		
12DL8	9HR	12DV8	12DS7
12DM4	**4CG**	12DM4A, 12DQ4	12D4, 12D4A
12DM4A	4CG	12DM4, 12DQ4	12D4, 12D4A
12DM5	7CV	12C5, 12CU5	12AS5, 12EH5, 12FX5
12DM7	9A	12DW7, 12DF7	12DT7, 12AX7, 12AX7A, 12AX7A/ECC83, 12AX7WA, 7025
12DQ4	4CG	12D4, 12D4A, 12DM4, 12DM4A	12AX4, 12AX4GT, 12AX4GTA, 12AX4GTB
12DQ6	6AM	12DQ6A, 12DQ6B, 12GW6	12CU6, 12BQ6, 12BQ6GA, 12BQ6GT, 12BQ6GTA, 12BQ6GTB
12DQ6A	6AM	12DQ6, 12DQ6B, 12GW6	12CU6, 12BQ6, 12BQ6GA, 12BQ6GT,

(continued)

Original	Base Diagram	Preferred Substitution	Good Substitution
			12BQ6GTA, 12BQ6GTB
12DQ6B	6AM	12DQ6, 12DQ6A, 12GW6	12CU6, 12BQ6, 12BQ6GA, 12BQ6GT, 12BQ6GTA, 12BQ6GTB
12DQ7	9BF	12BY7, 12BY7A, 12BV7	12GN7, 12GN7A
12DS7	9JU	12DS7A	
12DS7A	9JU	12DS7	
12DT5	9HN		
12DT7	9A	12AX7, 12AX7A, 12AX7A/ECC83, 12AX7WA, 7025, 12DF7, 12DM7	12AD7, 12BZ7, 12DW7
12DT8	9AJ		
12DU7	9JZ		
12DV8	9HR		12DL8
12DW4	9HP	12DW4A, 12CL3, 12BS3, 12BS3A	12B3, 12BK3
12DW4A	9HP	12DW4, 12CL3, 12BS3, 12BS3A	12B3, 12BK3
12DW5	9CK		12CM6
12DW7	9A	12AX7, 12AX7A, 12AX7WA, 12AX7A/ECC83, 7025, 12DF7, 12DM7, 12DT7	12AD7, 12BZ7
12DY8	9JD		
12DZ6	7BK	12EA6, 12EK6	
12DZ8	9JE		12FY8
12E5	6Q	12E5GT, 12J5, 12J5GT	
12E13	8ET	6550, 6CA7, 6CA7/EL34	
12EA6	7BK	12DZ6, 12EK6	12CX6

Original	Base Diagram	Preferred Substitution	Good Substitution
12EC8	9FA		
12ED5	7CV	12CA5, 12EH5, 12FX5	12AS5, 12C5, 12CU5, 12DM5
12EF6	7S		12EN6, 12W6
12EG6	7CH		12AD6, 12AG6, 12FA6, 12GA6
12EH5	7CV	12CA5, 12FX5, 12CU5, 12C5	12DM5
12EK6	7BK	12DZ6, 12EA6	12CX6
12EL6	7BF		
12EM6	9HV		
12EN6	7AC	12W6, 12W6GT, 12L6, 12L6GT	
12EQ7	9LQ		12KL8
12EX6	7BK	12EA6, 12EK6, 12DZ6	
12EZ6	7BK	12CY6	12AF6, 12BL6, 12EA6, 12EK6, 12DZ6
12F5GT	5M		
12F8	9FH		
12F31	7BK	12BA6	
12FA6	7CH		12AG6, 12GA6
12FB5	9CV		12ED5
12FK6	7BT	12FM6	12AE6
12FM6	7BT	12FK6	12AE6, 12AJ6, 12FT6
12FQ7	9LL		
12FQ8	9KT		
12FR8	9KU		
12FT6	7BT	12AE6, 12AE6A, 12BF6	12AJ6, 12FM6
12FV7	9A	12AV7	12AT7, 12AT7/ECC81, 12AT7WA, 12AT7WB, 12AZ7,

(continued)

Original	Base Diagram	Preferred Substitution	Good Substitution
			12AZ7A, 12DW7
12FX5	7CV	12ED5, 12EH5	12DM5, 12CA5, 12AS5
12FX8	9KV	12FX8A	
12FX8A	9KV	12FX8	
12G4	6BG	12H4	
12G11	12BU		12AL11
12GA6	7CH		12FA6, 12AD6, 12AG6
12GB3	6AM	12BQ6, 12BQ6GA, 12BQ6GT, 12BQ6GTA, 12BQ6GTB, 12CU6	12DQ6, 12DQ6A, 12DQ6B, 12GW6
12GB6	6AM	12DQ6, 12DQ6A, 12DQ6B, 12GW6	
12GB7	6AM	12DQ6, 12DQ6A, 12DQ6B, 12GW6	
12GC6	8JX		
12GE5	12BJ		
12GJ5	9QK	12GJ5A	
12GJ5A	9QK	12GJ5	
12GK17	4CG	12D4, 12D4A	
12GN7	9BF	12GN7A, 12HG7	
12GN7A	9BF	12GN7, 12HG7	
12GT5	9NZ	12GT5A	
12GT5A	9NZ	12GT5	
12GW6	6AM	6DQ6, 6DQ6A, 6DQ6B	
12H4	7DW		12G4
12H6	7Q		
12H31	7CH	12BE6	
12HE7	12FS		
12HG7	9BF	12GN7, 12GN7A	
12HL7	9BF	12GN7, 12GN7A	12HG7

Original	Base Diagram	Preferred Substitution	Good Substitution
12J5GT	6Q		
12J7	7R	12J7GT	12K7, 12K7GT
12J7GT	7R	12J7	12K7, 12K7GT
12J8	9GC		
12JB6	9QL	12JB6A	
12JB6A	9QL	12JB6	
12JF5			
12JN6	12FK		
12JN8	9FA		
12JQ6	9RA		
12JT6	9QU	12JT6A	
12JT6A	9QU	12JT6	
12K5	7EK		
12K7	7R	12K7GT	12J7, 12J7GT
12K7GT	7R	12K7	12J7, 12J7GT
12KL8	9LQ	12EQ7	
12L6	7AC	12L6GT, 12W6, 12W6GT	12EN6
12L6GT	7AC	12L6, 12W6, 12W6GT	12EN6
12L8	8BU		
12MD8	9RQ		
12Q7GT	7V		
12R5	7CV	12DM5, 12CU5, 12C5	12FX5, 12ED5, 12EH5, 12CA5
12RK19	9CB	12AF3, 12BR3	
12RLL3	9A	12AV7	
12RLL5	9LP	12FQ7	
12S8GT	8CB		
12SA7	8R	12SA7G, 12SA7GT, 12SA7GTY, 12SA7Y	12SY7
12SA7G	8R	12SA7, 12SA7GT, 12SA7GTY, 12SA7Y	12SY7

Original	Base Diagram	Preferred Substitution	Good Substitution
12SA7GT	8AD	12SA7, 12SA7G, 12SA7GTY, 12SA7Y	12SY7
12SA7GTY	8AD	12SA7, 12SA7G, 12SA7GT, 12SA7Y	12SY7
12SA7Y	8R	12SA7, 12SA7G, 12SA7GT, 12SA7GTY	12SY7
12SC7	8S		
12SF5	6AB	12SF5GT	
12SF5GT	6AB	12SF5	
12SF7	7AZ	12SF7GT, 12SF7Y	
12SF7GT	7AZ	12SF7, 12SF7Y	
12SF7Y	7AZ	12SF7, 12SF7GT	
12SG7	8BK	12SG7GT, 12SG7Y, 12SH7, 12SH7GT	
12SG7GT	8BK	12SG7, 12SG7Y, 12SH7, 12SH7GT	
12SG7Y	8BK	12SG7, 12SG7GT, 12SH7, 12SH7GT	
12SH7	8BK	12SH7GT, 12SG7, 12SG7GT, 12SG7Y	
12SH7GT	8BK	12SH7, 12SG7, 12SG7GT, 12SG7Y	
12SJ7	8N	12SJ7GT	12SK7, 12SK7G, 12SK7GT, 12SK7GTY, 12SK7Y
12SJ7GT	8N	12SJ7	12SK7, 12SK7G, 12SK7GT, 12SK7GTY, 12SK7Y
12SK7	8N	12SK7G,	12SJ7, 12SJ7GT

Original	Base Diagram	Preferred Substitution	Good Substitution
		12SK7GTY, 12SK7Y	
12SK7G	8N	12SK7, 12SK7GT, 12SK7GTY, 12SK7Y	12SJ7, 12SJ7GT
12SK7GT	8N	12SK7, 12SK7G, 12SK7GTY, 12SK7Y	12SJ7, 12SJ7GT
12SK7GTY	8N	12SK7, 12SK7G, 12SK7GT, 12SK7Y	12SJ7, 12SJ7GT
12SK7Y	8N	12SK7, 12SK7G, 12SK7GT, 12SK7GTY	12SJ7, 12SJ7GT
12SL7GT	8BD		
12SN7	8BD	12SN7GT, 12SN7GTA	12SX7, 12SX7GT
12SN7GT	8BD	12SN7, 12SN7GTA	12SX7, 12SX7GT
12SN7GTA	8BD	12SN7, 12SN7GT	12SX7, 12SX7GT
12SQ7	8Q	12SQ7G, 12SQ7GT, 12SR7, 12SR7GT	12SW7
12SQ7G	8Q	12SQ7, 12SQ7GT, 12SR7, 12SR7GT	12SW7
12SQ7GT	8Q	12SQ7, 12SQ7G, 12SR7, 12SR7GT	12SW7
12SR7	8Q	12SR7GT, 12SQ7, 12SQ7G, 12SQ7GT	12SW7
12SR7GT	8Q	12SR7, 12SQ7, 12SQ7G, 12SQ7GT	12SW7
12SW7	8Q		12SQ7, 12SQ7G, 12SQ7GT, 12SR7, 12SR7GT

Original	Base Diagram	Preferred Substitution	Good Substitution
12SX7	8BD	12SN7, 12SN7GT, 12SN7GTA	
12SY7	8R	12SA7, 12SA7G, 12SA7GT, 12SA7GTY, 12SA7Y	
12T10	12EZ		
12U7	9A		12AE7
12V6	7AC	12V6GT	12A6, 12EN6
12V6GT	7AC	12V6	12A6, 12EN6
12W6	7AC	12W6GT, 12L6, 12L6GT	12EN6
12W6GT	7AC	12W6, 12L6, 12L6GT	12EN6
12X4	5BS		
12Z3	4G		
13CW4	12AQ		
13D2	8BD	6SN7, 6SN7A, 6SN7GTA, 5SN7GTB, 6SN7GTY, 6SN7L	
13DE7	9HF		13DR7, 13FD7
13DR7	9HF	13FD7, 13FR7	
13EM7	8BD	15EA7	
13FD7	9HF	13FR7	
13FM7	12EJ	15FM7	
13FR7	9HF	13FD7	
13GB5	9NH	13GB5/XL500	
13GF7	9QD	13GF7A	
13GF7A	9QD	13GF7	
13J10	12BT	13Z10	
13JZ8	12DZ		
13V10	12EZ	12AE10	12BF11
14A4	5AC		

Original	Base Diagram	Preferred Substitution	Good Substitution
14A5	6AA		
14A7	8V	12B7	
14AF7	8AC	14N7	
14B6	8W		
14B8	8X		
14BL11	12GC		
14BR11	12GL		
14C5	6AA		
14C7	8V		
14E6	8W		
14E7	8AE		14R7
14F7	8AC		
14F8	8BW		
14GT8	9KR		14JG8
14H7	8V		14A7, 12B7
14J7	8BL	14S7	
14JG8	9KR		14GT8
14N7	8AC		12AF7
14Q7	8AL		
14R7	8AE		14E7
14S7	8BL		14J7
15	5F		
15AB9	10N		17AB9
15AE7	8BD	12EM7	
15AF11	12DP	15BD11, 15BD11A	
15BD11	12DP	15AF11, 15BD11A	
15BD11A	12DP	15BD11, 15AF11	
15CW5	9CV	15CW5/PL84	
15DQ8	9HX	15DQ8/PCL84	
15EA7	8BD	13EM7	
15EW7	9HF		
15FM7	12EJ	13FM7	

Original	Base Diagram	Preferred Substitution	Good Substitution
15FY7	12EO		
15HA6	9NW		15HB6
15HB6	9NW		15HA6
15KY8	9QT	15KY8A	15MX8
15KY8A	9QT	15KY8	15MX8
15LE8	9QZ		
15MF8	12DZ		
15MX8	9QT	15KY8, 15KY8A	
16A5	9CV	15CW5, 15CW5/PL84	
16A8	9EX	16A8/PCL82	
16AK9	12GZ		
16AQ3	9CB	16AQ3/XY88	
16BQ11	12DM		
16BX11	12CA		
16GK6	9GK		
16GY5	12DR		
16KA6	12GH		
16LU8	12DZ	16LU8A	17JZ8
16LU8A	12DZ	16LU8	17JZ8
16MY8	12DZ	16LU8, 16LU8A	17JZ8
17A8	9DC	19EA8	
17AB9	10N		15AB9
17AB10	12BT	17AX10, 17X10	
17AX3	12BL		
17AX4	4CG	17AX4GT, 17AX4GTA, 17D4	17DM4, 17DM4A, 17DQ5
17AX4GT	4CG	17AX4GTA, 17D4 17AX4GTA, 17D4	17DM4, 17DM4A, 17DQ4
17AX4GTA	4CG	17AX4, 17AX4GT, 17D4	17DM4, 17DM4A, 17DQ4
17AY3	9HP	17AY3A, 17BS3, 17BS3A	17CK3, 17CL3, 17DW4, 17DW4A
17AY3A	9HP	17AY3, 17BS3, 17BS3A	17CK3, 17CL3, 17DW4, 17DW4A

Original	Base Diagram	Preferred Substitution	Good Substitution
17BB14	9NH		
17BE3	12GA	17BZ3	
17BF11	12EZ		
17BH3	9HP	17BH3A	
17BH3A	9HP	17BH3	
17BQ6	6AM	17BQ6GTB	17GW6, 17DQ6B
17BQ6GTB	6AM	17BQ6	17GW6, 17DQ6B
17BR3	9CB	17RK19	
17BS3	9HP	17BS3A, 17DW4, 17DW4A, 17AY3, 17AY3A	17CK3, 17CL3
17BS3A	9HP	17BS3, 17DW4, 17DW4A, 17AY3, 17AY3A	17CK3, 17CL3
17BW3	12FX		
17BZ3	12GA	17BE3	
17C5	7CV	17CU5	
17C9	10F		
17CK3	9HP	17CL3, 17DW4, 17DW4A	17BS3, 17BS3A
17CL3	9HP	17CK3, 17BS3, 17BS3A, 17DW4, 17DW4A	
17CQ4	4CG	17DE4	
17CT3	9RX		
17CU5	7CV	17C5	17R5
17D4	4CG	17DM4, 17DM4A, 17DQ4, 17AX4, 17AX4GT, 17AX4GTA	
17DE4	4CG	17CQ4	
17DM4	4CG	17DM4A, 17DQ4	17D4
17DM4A	4CG	17DM4, 17DQ4	17D4
17DQ4	4CG	17DM4, 17DM4A	17D4

Original	Base Diagram	Preferred Substitution	Good Substitution
17DQ6	6AM	17DQ6A, 17DQ6B, 17GW6	
17DQ6A	6AM	17DQ6, 17DQ6B, 17GW6	
17DQ6B	6AM	17DQ6, 17DQ6A, 17GW6	
17DW4	9HP	17DW4A, 17BS3, 17BS3A	17CK3, 17CL3
17DW4A	9HP	17DW4, 17BS3, 17BS3A	17CK3, 17CL3
17EW8	9AJ	17EW8/HCC85	17JK8
17GE5	12BJ		
17GJ5	9QK	17GJ5A	
17GJ5A	9QK	17GJ5	
17GT5	9NZ	17GT5A	
17GT5A	9NZ	17GT5	
17GV5	12DR		
17GW6	6AM	17DQ6, 17DQ6A, 17DQ6B	
17H3	9FK		
17HB25	17HB25		
17JB6	9QL	17JB6A	
17JB6A	9QL	17JB6	
17JG6	9QU	17JG6A, 17JT6, 17JT6A	17JR6, 17KV6, 17KV6A
17JG6A	9QU	17JG6, 17JT6, 17JT6A	17JR6, 17KV6, 17KV6A
17JM6	12FJ	17JM6A	
17JM6A	12FJ	17JM6	
17JN6	12FK		
17JQ6	9RA		
17JR6	9QU	17JG6, 17JG6A, 17KV6, 17KV6A	
17JT6	9QU	17JT6A, 17JG6, 17JG6A	17JR6, 17KV6, 17KV6A

Original	Base Diagram	Preferred Substitution	Good Substitution
17JT6A	9QU	17JT6, 17JG6, 17JG6A	17JR6, 17KV6, 17KV6A
17JZ8	12DZ		
17KV6	9QU	17KV6A, 17JR6	17JG6, 17JG6A
17KV6A	9QU	17KV6, 17JR6	17JG6, 17JG6A
17L6	7AC	17W6	
17LD8	9QT	15KY8, 15KY8A	
17R5	7CV		17C5, 17CA5, 17CU5
17RK19	9CB	17BR3	
17W6	7AC	17L6	
17X10	12BT	17AB10	
17Y9	10L		
17Z3	9CB	17Z3/PL81	
18A5	6CK		
18AJ10	12EZ		
18FW6	7CC	18FW6A	18GD6, 18GD6A
18FW6A	7CC	18FW6	18GD6, 18GD6A
18FX6	7CH	18FX6A	
18FX6A	7CH	18FX6	
18FY6	7BT	18FY6A	18GE6, 18GE6A
18FY6A	7BT	18FY6	18GE6, 18GE6A
18GB5	9NH	18GB5/LL500	
18GD6	7BK	18GD6A	
18GD6A	7BK	18GD6	
18GE6	7BT	18GE6A, 18FY6A	
18GE6A	7BT	18GE6, 18FY6, 18FY6A	
18GV8	9LY	18GV8/PCL85	
19	6C		
19AU4	4CG	19AU4GTA	
19AU4GTA	4CG	19AU4	
19BG6G	5BT	19BG6GA	
19BG6GA	5BT	19BG6G	

Original	Base Diagram	Preferred Substitution	Good Substitution
19C8	9E		19T8
19CG3	12HF	19DQ3	
19CL8A	9FX	19JN8	
19DE3	12HX		
19DE7	9HF		19EW7
19DK3	9SG		
19DQ3	12HF	19CG3	
19EA8	9AE		17A8
19EZ8	9KA		
19FX5	7CV		
19GQ7	9QM		
19HR6	7BK		19HS6
19HS6	7BK		19HR6
19HV8	9FA		
19J6	7BF		
19JN8	9FA	19CL8, 19CL8A	
19KG8	9LY		
19MR9	7BK	18GD6, 18GD6A	
19MR10	7BK	18GD6, 18GD6A	
19MR19	7CC	18FW6, 18FW6A	
19Q9	10H		
19T8	9E		19C8
19X8	9AK		
20	4D		
20A3	7BN	2D21, 2D21W	
20AQ3	9CB	20AQ3/LY88	
20EQ7	9LQ		
20EW7	9HF		19EW7
20EZ7	9PG		
20LF6	12GW		
21EX6	5BT	25CD6, 25CD6GA, 25CD6GB, 35DN6	
21GY5	12DR		
21HB5	12BJ	21HB5A	

Original	Base Diagram	Preferred Substitution	Good Substitution
21HB5A	12BJ	21HB5	
21HJ5	12FL		
21JS6A	12FY	23JS6A	
21JV6	12FK		
21JZ6	12GD		
21KA6	12GH		
21KQ6	9RJ		
21LG6	12HL	21LG6A	
21LG6A	12HL	21LG6	
21LR8	9QT		
21LU8	12DZ		21MY8
21MY8	12DZ	21LU8	
22	4K		
22BH3	9HP	22BH3A	
22BH3A	9HP	22BH3	
22BW3	12FX		
22DE4	4CG		
22JF6	9QL	22KM6	22JU6
22JG6	9QU	22JG6A	22JR6
22JG6A	9QU	22JG6	22JR6
22JR6	9QU		22JG6, 22JG6A
22JU6	9QL	22KM6	22JF6
22KM6	9QL		22JU6
23JS6A	12FY		21JS6A
23Z9	12GZ		
24A	5E	35, 35/51	
24BF11	12EZ		
24JE6	9QL	24JE6A, 24JE6B, 24JE6C, 24LQ6	
24JE6A	9QL	24JE6, 24JE6B, 24JE6C, 24LQ6	
24JE6B	9QL	24JE6, 24JE6A, 24JE6C, 24LQ6	
24JE6C	9QL	24JE6, 24JE6A, 24JE6B, 24LQ6	

Original	Base Diagram	Preferred Substitution	Good Substitution
24JZ8	12DZ		25JZ8
24LQ6	9QL	24JE6, 24JE6A, 24JE6B, 24JE6C	
24LZ6	9QL		
25A6	7S	25A6GT	25W6, 25W6GT, 25L6, 25L6GT
25A6GT	7S	25A6	25L6, 25L6GT, 25W6, 25W6GT
25A7GT	8F		
25AC5GT	6Q		
25AV5GA	6CK		
25AX4	4CG	25AX4GT, 25D4	25W4, 25W4GT
25AX4GT	4CG	25AX4, 25D4	25W4, 25W4GT
25B5	6D		
25B6	7S	25B6G, 25C6, 25C6G, 5824	
25B6G	7S	25B6, 25C6, 25C6G, 5824	
25B8GT	8T		
25BK5	9BQ		
25BQ6	6AM	25BQ6GA, 25BQ6GT, 25BQ6GTB, 25CU6	25DQ6
25BQ6GA	6AM	25BQ6, 25BQ6GT, 25BQ6GTB, 25CU6	25DQ6
25BQ6GT	6AM	25BQ6, 25BQ6GA, 25BQ6GTB, 25CU6	25DQ6
25BQ6GTB	6AM	25BQ6, 25BQ6GA, 25BQ6GT, 25CU6	25DQ6
25C5	7CV		25CA5
25C6G	7AC		
25CA5	7CV	25C5	25EH5
25CD6	5BT	25CD6GA, 25CD6GB, 25DN6	
25CD6GA	5BT	25CD6, 25CD6GB, 25DN6	

Original	Base Diagram	Preferred Substitution	Good Substitution
25CD6GB	5BT	25CD6, 25CD6GA, 25DN6	
25CG3	12HF		
25CK3	9HP		25CM3
25CM3	9HP		25CK3
25CT3	9RX		
25CU6	6AM	25BQ6, 25BQ6GA, 25BQ6GT, 25BQ6GTB	
25DL3	9GD		
25DN6	5BT	25CD6, 25CD6GA, 25CD6GB	25EC6
25E5	8GT	25E5/PL36	
25EC6	5BT	25CD6, 25CD6GA, 25CD6GB	25DN6
25EH5	7CV	25CA5	25C5
25F5A	7CV		25C5, 25CA5
25GB6	6AM	25BQ6, 25BQ6GA, 25BQ6GT, 25BQ6GTB, 25CU6	
25HX5	9SB		
25JQ6	9RA		
25JZ8	12DZ		24JZ8
25L6	7AC	25L6G, 25L6GT, 25W6, 25W6GT	25C6G
25L6G	7AC	25L6, 25L6GT, 25W6, 25W6GT	25C6G
35L6GT	7AC	25L6, 25L6G, 25W6, 25W6GT	25C6
25N6G	7W		
25T			
25W4	4CG	25W6GT, 25AX4, 25AX4GT	25D4
25W4GT	4CG	25W4, 25AX4, 25AX4GT	25D4

Original	Base Diagram	Preferred Substitution	Good Substitution
25W6	7AC	25W6GT, 25L6, 25L6G, 25L6GT	
25W6GT	7AC	25W6, 25L6, 25L6G, 25L6GT	
25X6	7Q	25Z6, 25Z6G, 25Z6GT, 25Z6MG	
25Y5	6E		25Z5
25Z5	6E		
25Z6	7Q	25Z6G, 25Z6GT, 25Z6MG	
25Z6G	7Q	25Z6, 25Z6GT, 25Z6MG	
25Z6GT	7Q	25Z6, 25Z6G, 25Z6MG	
25Z6MG	7Q	25Z6, 25Z6G, 25Z6GT	
26	4D		
26A6	7BK	26CG6	
26A7GT	8BU		
26C6	7BT		
26CG6	7BK	26A6	
26D6	7CH		
26HU5	8NB		
26LW6	8NC		
26LX6	12JA		
27	5A		56
27GB5	9NH	27GB5/PL500, 28GB5	
28GB5	9NH	27GB5, 27GB5/PL500	
29KQ6	9RJ	29KQ6/PL521	29LE6
29LE6	9RL		29KQ6, 29KQ6/PL521
30	4D		

Original	Base Diagram	Preferred Substitution	Good Substitution
30A5	7CV	35C5, 35EH5	
30AE3	9CB	30AE3/PY88	
20AG11	12DA		
30C1	9DC	9A8, 9A8, PCF80	
30JZ6	12GD		
30KD6	12GW		
30MB6	12FY		
30P4	8GT	25E5, 25E5/PL36	
30P18	9CV	15CW5, 15CW5/PL84	
30P19	8GT	25E5, 25E5/PL36	
30PL12	9EX	16A8, 16A8/PCL82	
30PL13	9GK	16GK6	
20PL14	9GK	16GK6	
31	4D		
31AL10	12HR		
31JS6	12FY	31JS6A, 31JS6C	
31JS6A	12FY	31JS6, 31JS6C	
31JS6C	12FY	31JS6, 31JS6A	
31LQ6	9QL		
31LR8	9QT		
31LZ6	9QL	35MC6	
32	4K	1A4, 34	1B4, 1B4P
32ET5	7CV	32ET5A, 24GD5, 34GD5A	
32ET5A	7CV	32ET5, 34GD5, 34GD5A	
32HQ7	12HT		
32L7GT	8Z		
33	5K		
33GT7	12FC		
33GY7	12FN	33GY7A	

Original	Base Diagram	Preferred Substitution	Good Substitution
33JR6	9QU		
33JV6	12FK		
34	4M	1A4	1B4, 1B4P, 32
34CE3	12GK		
34CM3	9HP		
34GD5	7CV	34GD5A	32ET5, 32ET5A
34GD5A	7CV	34GD5	32ET5, 32ET5A
34R3	9CB		
35	5E	51	24A
35A5	6AA		
35B5	7BZ		
35C5	7CV		
35DZ8	9JE		
35EH5	7CV		
35GL6	7FZ		
35L6GT	7AC		
35LR6	12FY		
35LR6		2-5DA	
35T			
35TG			
35W4	5BQ		
35Y4	5AL		
35Z3	4Z		
35Z4GT	5AA		
35Z5	6AD	35Z5G, 35Z5GT	
35Z5G	6AD	35Z5, 35Z5GT	
35Z5GT	6AD	35Z5, 35Z5G	
36	5E		
36AM3	5BQ	36AM3A, 36AM3B	
36AM3A	5BQ	36AM3, 36AM3B	
36AM3B	5BQ	36AM3, 36AM3A	
36KD6	12GW	40KD6	
36MC6	9QL	31LZ6	
37	5A	76	

Original	Base Diagram	Preferred Substitution	Good Substitution
38	5F		
38HE7	12FS		38HK7
38HK7	12FS	38HE7	
39	5F	39/44	44
40	4D		
40KD6	12GW	36KD6/40KD6	36KD6
40KG6	9RJ	40KG6A, 40KG6A/PL509	
40KG6A	9RJ	40KG6, 40KG6A/PL509	
40Z5	6AD		45Z5, 45Z5GT
41	6B		42
42	6B		41
42EC4	6EC4	42EC4A, 42EC4A/PY500	
42EC4A	6EC4	42EC4, 42EC4A/PY500	
42KN6	12GU		
42	6B		
44	5F	39/44	39
45	4D		
45Z3	5AM		
45Z5GT	6AD		
46	5C		
47	5B		
48	6A		
48A8	9EX	50BM8, 50BM8/UCL82	
49	5C		
50	4D		
50A5	6AA		
50B5	7BZ		
50BM8	9EX	50BM8/UCL82	48A8
50C5	7CV		
50C6G	7AC		

Original	Base Diagram	Preferred Substitution	Good Substitution
50CA5	7CV		50C5, 50EH5
50DC4	5BQ		
50EH5	7CV	50CA5	50C5
50FA5	7CV		50FK5
50FK5	7CV		50FA5
50GY7A	12FN		
50HC6	7FZ		50HK6
50HK6	7FZ		50HC6
50JY6	8MG		
50L6GT	7AC		
50Y7GT	8AN		
50Z6	7Q		50AX6
50Z7	8AN	50Z7G, 50Y7, 50Y7GT	
51	5E	35	
52KU	5T	5Y3, 5Y3G, 5Y3GA, 5Y3GT	
53	7B		
53HK7	12FS		58HE7
53KU	5T	5Y3, 5Y3G, 5Y3GA, 5Y3GT	
54KU	5T	5Y3, 5Y3G, 5Y3GA, 5Y3GT	
56	5A		27
56R9	12EN		
57	6F		58
58	6F		
58HE7	12FS		53HK7
60FX5	7CV		
64	5E		65, 36
65	5E		36
70L7GT	8AA		
75	6G		
75TH			
75TL			

Original	Base Diagram	Preferred Substitution	Good Substitution
76	5A		37
77	6F	6C6	57
78	6F	6D6	58
80	4C		5Z3, 83
83	4C	5Z3	
81	4B		
81-A	3G		
82	4C		
83	4C	5Z3	
84	5D	6Z4	
85	6G		
88	4C		82
89	6F		
100R		8020	
100TH			
100TL			
108C1	5BO	0B2, 0B2WA	0C3, 0C3A
117L7	8AO	117L7GT, 117M7, 117M7GT	117P7, 117N7, 117N7GT, 117P7GT
117L7GT	8AO	117L7, 117L7, 117M7, 117M7GT	117P7, 117N7, 117P7GT, 117N7GT
117M7	8AO	117M7GT, 117L7, 117L7GT	
117M7GT	8AO	117M7, 117L7, 117L7GT	
117N7	8AV	117N7GT, 117P7, 117P7GT	
117N7GT	8AV	117N7, 117P7, 117P7GT	
117P7	8AV	117P7GT, 117N7, 117N7GT	
117Z3	4CB		
117Z4	5AA	117Z4GT	
117Z4GT	5AA	117Z4	

Original	Base Diagram	Preferred Substitution	Good Substitution
117Z6	7Q	117Z6GT	
117Z6GT	7Q	117Z6	
182/482	4D		183/483
183/483	4D		182/482
150C1	5BO	0A2, 0A2WA, 6073, 6626	
150C2	5BO	0A2, 0A2WA, 6073, 6626	
150C3	4AJ	0D3, 0D3A	
150C4	5BO	0A2WA, 6073, 6626	0A2
152TH			
152TL			
175A			
177A			
177WA			
		2-150D	
180C1	5BO	0B2, 0B2WA, 6074	
245	6Q2	884	
250R			
250TH			
250TL			
253			
254W			
264		8576	
274	5L	5V4, 5V4G, 5V4GA, 5R4 5R4GB, 5R4GY, 5R4GYB 274A, 274B	
274A	5L	274, 274B, 5V4, 5V4G, 5V4GA, 5R4, 5R4G, 5R4GB, 5R4GY, 5R4GYB	
274B	5L	274, 274A, 5V4, 5V4G, 5V4GA, 5R4,	

132

Original	Base Diagram	Preferred Substitution	Good Substitution
		5R4G. 5R4GB 5R4GY. 5R4GYB	
279			
284			
290			
290A			
294			
301	4C	301A. 83	
301A	4C	301. 83	
304TH			
304TL			
310	7R	310B. 348	1620
310B	7R	310. 348	1620
313	4V	1C21. 313C	
313C	4V	313. 1C21	
322			
328A	6F		6C6
348	7R	348A	1620
348A	7R	348	1620
349A	7S	6K6. 6K6G. 6K6GT. 6K6MG. 6F6. 6F6G. 6F6GT. 6F6MG	
351ᴧ	6S	6X5. 6X5G. 6X5GT. 6X5L. 6X5MG. 6X5W	
359	4V	359A. 1C21	
359A	4V	359. 1C21	
381		7289	
395	4CK	5823. 395A	
395A	4CK	395. 5823	
401	7BD	401B. 5590	
401B	7BD	401. 5590	
403	7BD	403A. 403B. 6AK5. 6AK5/EF95. 5654	

Original	Base Diagram	Preferred Substitution	Good Substitution
403A	7BD	403, 403B, 6AK5, 6AK5/EF95, 5654	
403B	7BD	403, 403A, 6AK5, 6AK5/EF95, 5654	
404	9X	404A, 5847	
404A	9X	404, 5847	
407A	7BH		2C51
408A	7BD	6AK5, 6AK5/EF95	
409	7CM	409A, 6AS6	
409A	7CM	409, 6AS6	
417	9V	5842, 417A	
417A	9V	417, 5842	
421	8BD	6AS7, 6AS7G, 6AS7GA, 6080, 421A	
421A	8BD	421, 6AS7, 6AS7G, 6AS7GA, 6080	
423	5BO	423A, 5651, 5651WA	
423A	5BO	423, 5651, 5651WA	
450TH			
450TL			
451		8020, 100R	
502A	6BS	2050, 2050A	
546	7BN	5696, 5696A	
592		3-200A3	
630	6BS	2050, 2050A, 630A	
630A	6BS	630, 2050, 2050A	
731A	7BD	6AK5/EF95	

Original	Base Diagram	Preferred Substitution	Good Substitution
750TL			
807	5AW		5933
826			
866	4P		3B28
879	4AB		2X2
884	6Q2	6Q5	885
885	6Q2	884, 2B4	
950	5K		1F4
954	5BB		956, 9001
955	5BC		5731
956	5BC		5731, 9003
958A	7BS		9002
959	5BE		
991	991		
1000T			
1201	8BN		7E5
1203	4AH		7C4
1204	8BO		7AB7
1205	4AJ		0A3, 0A3A
1206	4AJ		0C3, 0C3A
1207	5BO		0A2, 0A2WA
1208	5BO		0B2, 0B2WA
1210	5BO		0A2WA
1211	5BO		0B2WA
1217	7CH	6BE6, 5915	
1219	8CJ	5670	
1221	6F		6C6
1223	7R		1620, 6J7, 6J7G, 6J7GT, 6J7GTX, 6J7MG
1225	8B		6L7
1229	4K	1B4, 1B4P	1A4, 1A4P, 1A4T
1231	8V	7V7	
1232	8V		7G7

Original	Base Diagram	Preferred Substitution	Good Substitution
1266	4CK		5823
1267	4V	0A4G	
1273	8V	7A7	7AJ7
1275	4C		5Z3, 5X3
1280	8V		12C7
1291	7BE		3B7, 1288, 1292
1294	4AH		1R4
1299	6BB		3D6
1381HQ	7BD	6AK5/EF95, 5654	
1500T			
1603	6F		6C6, 1221
1611	7S	1621	6F6, 6F6GT
1612	7T		6L7
1613	7S	1621	6F6, 6F6GT
1614	7AC	6L6, 6L6G, 6L6GA, 6L6GB, 6L6GC, 6L6GT, 6L6GX, 6L6Y, 6L6WA, 6L6WB, 6L6WGT	1631, 5881, 7581
1620	7R	6J7, 6W7, 6J7G, 6J7GT, 6J7GTX 6J7MG	1223, 7000
1621	7S	6F6, 6F6G, 6F6GT, 6F6MG, 6K6, 6K6G, 6K6GT, 6K6MG	
1622	7AC	6L6, 6L6G, 6L6GA, 6L6GB, 6L6GC, 6L6GT, 6L6GX, 6L6Y, 6L6WA, 6L6WB, 6L6WGT	5932, 5881
1629	6R	6E5	
1631	7AC	1614, 1622, 6L6, 6L6G, 6L6GA,	

Original	Base Diagram	Preferred Substitution	Good Substitution
		6L6GB, 6L6GC, 6L6GT, 6L6GX 6L6Y, 6L6WA, 6L6WB, 6L6BGT	
1632	7AC		12L6, 12L6GT, 12W6, 12S6GT
1634	8S		12SC7
1635	8B		6Y7, 6Y7G, 6N7, 6N7G, 6N7GT, 6N7MG
1642	7BH	2C21	
1644	8BU		12L8
1649	8N		6AC7, 6AC7W
1650	5BC	955	
1655	8S		6SC7, 6SC7GT, 6SC7GTY
1657	6BS		2050, 2050A
1665	6BS		2030, 2050A
1852	8N		6AC7, 6AC7W
1853	8N		6AB7
2000R		2-2000A	
2000T			
2013	9A	6211	
2014	9BV	6197	6CL6, 6677
2050	6BS	2050A	
2050A	6BS	2050	
2057	7Q	2057/6H6	6H6, 6H6G, 6H6GT, 6H6MG
2076	5T	5R4, 5R4G, 5R4GB, 5R4GY, 2076/5R4GB, 2076/5R4GYB, 5R4GYB	
2081	9DX	2081/6AW8A	6AW8, 6AW8A
2082	9A	2082/12AY7	12AY7
2100		8020, 100R	

Original	Base Diagram	Preferred Substitution	Good Substitution
2101	5K		1F4
3107	5L	5V4, 5V4G, 5V4GA	
3861B		7034, 4X150A	
4707	5BS	6X4, 6X4W	
5559	4BL	5720, 5728	
5590	7BD	6AJ5, 6AK5, 6AK5/EF95	5654, 5591
5591	7BD	6AJ5, 6AK5, 6AK5/EF95	5654, 5590
5633		5634	
5634		5633	
5636	8DC	5636A	6205, 6206, 8517, 8522, 8524
5636A	8DC	5636	6205, 6206, 8517, 8522, 8524
5639	8DL	8211	
5642			
5651	5BO	5651A, 5641WA	423, 423A
5651A	5BO	5651, 5651WA	423, 423A
5651WA	5BO	5651, 5651A	423, 423A
5654	7BD	6AK5, 6AK5/EF95, 5654/6AK5W 6096, 5654W	
5654W	7BD	5654, 6AK5, 6AK5/EF95, 5654/6AK5W 6096	
5659	7S		12A6
5660	8E		12C8
5661	8N		12SK7, 12SK7G, 12SK7GT, 12SK7GTY,

Original	Base Diagram	Preferred Substitution	Good Substitution
			12SK7Y
5663	6CE		5696, 5696A
5670	8CJ	5670WA, 6385, 6854	6AK5, 6AK5/EF95
5670WA	8CJ	5670, 6385, 6854	6AK5, 6AK5/EF95
5672	5672		
5678	5678		
5679	7AJ		7A6
5686	9G		
5687	9H	7044	
5691	9BD	6113, 6188	6SL7, 6SL7A, 6SL7GT, 6SL7GTY, 6SL7L
5692	8BD		6SN7, 6SN7A, 6SN7GTA, 6SN7GTB, 6SN7GTY, 6SN7L
5693	8N		6SJ7, 6SJ7GT, 6SJ7GTX, 6SJ7GTY, 6SJ7Y
5696	7BN	5696A	
5696A	7BN	5696	
5718	8DK	5897	
5719	8DK		
5720	4BL	5559, 5728	
5725	7CM	6AS6, 6AS6W	6187
5726	6BT	6AL5, 6097, 6663, 6AL5W	
5727	7BN	2D21, 2D21W	
5728	4BL	5559, 5720	
5731	5BC	955	
5732	7R	6K7, 6K7G, 6K7GT, 6K7GTX, 6K7MG	
5734	5734		

Original	Base Diagram	Preferred Substitution	Good Substitution
5749	7BK	6BA6, 6BA6/EF93, 6BA6W, 6660, 8425	
5750	7CH		6BE6
5751	9A	5751WA	6681, 12AX7, 12AX7A, 12AX7WA, 12AX7A/ECC83
5751WA	9A	5751	6681, 12AX7, 12AX7A, 12AX7WA, 12AX7A/ECC83
5755	8BD		6SU7
5763	9K	6062	
5783	5783		
5814	9A	5814A, 5814WA, 6680, 7247, 7728, 7730	12AU7A/ECC82, 6189
5814A	9A	5814, 5814WA, 6680, 7247, 7728, 7730	12AU7A/ECC82, 6189
5814WA	9A	5814, 5814A, 6680, 7247, 7728, 7730	12AU7A/ECC82, 6189
5823	4CK		
5824	7S		25B6
5840	8DE	5840A, 5840W, 5899, 8529	
5840A	8DE	5840, 5840W, 5899, 8529	
5840W	8DE	5840, 5840A, 5899, 8529	
5842	9V	417, 417A	
5844	7BF		5964, 6J6, 6J6A, 6J6WA, 6J6WB
5847	9X	404, 404A	
5867A			

Original	Base Diagram	Preferred Substitution	Good Substitution
5871	9X		6V6, 6V6G, 6V6GT, 6V6GTA, 6V6GTX, 6V6GTY, 6V6GX, 6V6Y
5879	9AD		
5881	7AC		6L6, 6L6G, 6L6GA, 6L6GB, 6L6GC, 6L6GT, 6L6GX, 6L6Y
5896	8DJ		
5897	8DK	5718, 8527	
5899	8DE	5899A, 5840, 8529, 5840W	
5899A	8DE	5899, 5840, 8527, 5840W	
5900	8DL	5901, 5899, 5899A, 8529	
5901	8DL	8530, 5840, 5840W	
5902	8DE	8528	
5906	8DL	8064	
5908	8DC	8414	
5910	6AR		1U4
5915	7CH	5915A 5750, 7036	6BE6
5915A	7CH	5915, 5750, 7036	6BE6
5920	7BF		5963, 6101, 6J6, 6J6A, 6J6WA, 6J6WB
5930	4D		2A3
5931	5T		5U4, 5U4G, 5U4GB
5932	7AC	7581, 1622, 7027, 7027A	6L6, 6L6A, 6L6G, 6L6GA, 6L6GB, 6L6GC, 6L6GT, 6L6GX, 6L6Y

Original	Base Diagram	Preferred Substitution	Good Substitution
5933	5AW	807	
5961	8R		6SA7, 6SA7G, 6SA7GT, 6SA7GTX 6SA7GTY, 6SA7Y
5963	9A		12AU7, 12AU7A, 12AU7A/ECC82, 6680
5964	7BF	6101, 6045, 7244, 6030	6J6, 6J6A, 6J6WA, 6J6WB
5965	9A	5965A, 6829	
5965A	9A	5965, 6829	
5992	7AC	5871, 6V6, 6V6G, 6V6GT, 6V6GTA, 6V6GTX, 6V6GTY, 6V6GX, 6V6Y	
5998	8BD	6080, 6520, 7236, 7802	
6005	7BZ	6095, 6669	6AQ5, 6AQ5A, 6AQ5W
6006	8BK	6SG7, 6SG7GT, 6SG7Y	
6012	7BN	5727	2D21
6021	8DG		
6028	7BD	408A	
6030	7BF	5964, 6099, 6101	6J6, 6J6A, 6J6WA, 6J6WB
6045	7BF	5964, 6101, 6099	6J6, 6J6A, 6J6WA, 6J6WB
6046	7AC	5824	25L6, 25L6G, 25L6GT
6049	8DL	5840, 5900, 5901, 6225	
6057	9A	6681	12AX7, 12AX7A, 12AX7A/ECC83
6058	6BT	5726, 6097, 6663	6AL5

Original	Base Diagram	Preferred Substitution	Good Substitution
6060	9A	6201, 6679	12AT7, 12AT7/ECC81, 12AT7WA, 12AT7WB
6061	9AM	6BW6	
6062	9K		5763
6063	5BS	6202	6X4, 6X4W
6064	7BD	7498	6AM6, 6AM6/EF91
6065	7BK		6CQ6
6066	7BT		6AT6
6067	9A	5814, 5814A, 5814WA, 12AU7, 12AU7A, 6680, 12AU7A/ECC82	
6072	9A	6072A, 7247, 2082, 12AY7	
6072A	9A	6072, 7247, 2082, 12AY7	
6073	5BO	6626, 0A2, 0A2WA	
6074	5BO	6627, 0B2, 0B2WA	
6080	8BD	5998, 6520, 7802, 6080WA	6AS7, 6AS7G, 6AS7GA
6080WA	8BD	6080, 5998, 6520, 7802	6AS7, 6AS7G, 6AS7GA
6082	8BD	6082A	
6082A	8BD	6082	
6084	9AD	5879	
6085	8BD		6SN7, 6SN7A, 6SN7GTA, 6SN7GTB, 6SN7GTY, 6SN7L
6087	5L	6106, 6853	5Y3, 5Y3G, 5Y3GA, 5Y3GT
6092	2E31	5672	

Original	Base Diagram	Preferred Substitution	Good Substitution
6094	7BZ	6005. 6AQ5. 6AQ5A	
6095	7BZ	6005. 6669. 6094	6AQ5. 6AQ5A
6096	7BD	5591. 5664. 6968	5654. 6AK5. 6AK5/EF95
6097	6BT	5726. 6663. 6058	6AL5
6098	6BQ	6384	6AR6
6099	7BF	5964. 6101	6J6. 6J6A. 6J6WA. 6J6WB
6100	6BG	6135	6C4
6101	7BF	5964. 6099. 7244	6J6. 6J6A. 6J6WA. 6J6WB
6106	5T		6853. 5Y3. 5Y3G. 5Y3GA. 5Y3GT
6111	8DG	7079. 8526	
6112	8DG		
6113	8BD	6188. 5691	6SL7. 6SL7A. 6SL7GT. 6SL7GTY. 6SL7L
6132	6F	7499	6CH6
6134	8N	1852	6AC7. 6AC7W
6135	6BG	6100	6C4
6136	7BK	6660. 7543. 8425	6AU6. 6AU6A
6137	8N		6SK7. 6SK7G. 6SK7GT. 6SK7GTX. 6SK7GTY. 6SK7Y
6140	5BO		5651. 5651A. 5651WA. 423. 423A
6146	7CK	8298A	
6147	6CL	6397	

144

Original	Base Diagram	Preferred Substitution	Good Substitution
6152		5975	
6155			
6156			
6159	7CK	7357	
6180	8BD	5692	6SN7, 6SN7A, 6SN7GTA, 6SN7GTB, 6SN7GTY, 6SN7L
6185	8CJ	2C51, 1219, 5670	
6186	7BD	6AG5, 6186W	
6186W	7BD	6186, 6AG5	
6187	7CM	6AS6, 5725	
6188	8BD	6113, 5691	6SL7, 6SL7A, 6SL7GT, 6SL7GTY, 6SL7L
6189	9A	5814, 5814A, 6679, 6680, 7247, 7730, 5963	12AU7, 12AU7A, 12AU7WA, 12AU7A/ECC82
6197	9BV	6CL6, 6677	
6201	9A	6679, 12AT7, 12AT7WA, 12AT7WB	
6202	5BS	6X4, 6X4W, 6063	
6205	8DC	6206, 6943, 8444	
6206	8DC	6944, 5899	
6211	9A	6211A	6J6, 6J6A, 6J6WA, 6J6WB
6211A	9A	6211	6J6, 6J6A, 6J6WA, 6J6WB
6223	8DL	5840, 5901	
6224	8DL	5902	
6225	8DL	5899, 5900	
6245	5702	6540, 7083	

Original	Base Diagram	Preferred Substitution	Good Substitution
6265	7CM	6661, 7732	6BH6
6320	8DG	6112	
6321	8DG	6320, 6112	
6336	8BD	6336A, 6528, 6337	
6336A	8BD	6226, 6337, 6528	
6337	8BD	6336, 6336A	
6350	9CZ		
6360	9PW	6360A, 8457	
6360A	9PW	6360, 8457	
6385	8CJ	2C51, 6854, 5670	
6386	8CJ		2C51
6394	8BD	6083	
6397	6CL	6147	
6414	9A	6829, 5965	12AV7
6417	9K		5763, 7551
6418	512AX	6519	
6463	9CZ	6840	
6485	7BK		6AH6
6486	7CM	6486A	5725, 6AS6
6486A	7CM	6486	5725, 6AS6
6519	512AX	6418	
6520	8BD	6080, 7802	6AS7, 6AS7G, 6AS7GA
6528	8BD		6080
6540	5702	6872, 7083, 5702, 6245	
6549			
6549W		177WA	
6550	7S		7027, 7027A
6569			
6580			

146

Original	Base Diagram	Preferred Substitution	Good Substitution
6626	5BO	0A2, 0A2WA, 6073	150C4
6627	5BO	0B2, 0B2WA, 6074	108C1
6660	7BK	6BA6, 6BA6/EF93, 5749, 6136, 7543, 8425	
6661	7CM	6275, 7693, 6BH6	
6662	7CM	6BJ6	
6663	6BT	5726, 6098, 6AL5	
6664	5CE	6AB4	
6669	7BZ	6005, 6095, 6AQ5, 6AQ5A	
6676	7CM	7732, 6CB6A	6CB6, 6CF6, 6BZ6
6677	9BV	6CL6, 6197	
6678	9AE	6U8, 6U8A	7687, 7731
6679	9A	6201, 6060, 12AT7, 12AT7WA, 12AT7WB	
6680	9A	12AU7, 12AU7A, 6189, 7247, 7318, 12AU7A/ECC82	
6681	9A	12AX7, 12AX7A, 12AX7A/ECC83, 7025, 7247, 6057	
6686	9AU		
6687	7CH	5915	
6688A	9EQ		
6696A			
6697A			
6775		4-400C	
6807		6809, 6858	

Original	Base Diagram	Preferred Substitution	Good Substitution
6808		6859	
6809		6807, 6858	
6816			
6829	9A	5965	
6840	9CZ	6463	
6851	9A		6414
6853	5T	5Y3, 5Y3G, 5Y3GA, 5Y3GT	6087, 6106
6854	8CJ	2C51, 5670, 6385	
6858		6807, 6809	
6859		6808	
6872	5702	6540, 7083, 5702, 6245	
6883	7CK	8032, 8552	
6884			
6887	6BT	6919	6AL5
6894			
6895			
6919	6BT	6887	
6922	9AJ	7803, 6922/E88CC	
6939	9HL		
6943	8DC		6944
6944	8DC		
6948	8DG	6112	
6954	7CM		6DB6
6955	9A	7318	
6968	7BD	5654, 6096, 6AK5, 6AK5/EF95	
6973	9EU		
6977	6977		
7000	7R	1620, 6054	6J7, 6J7G, 6J7GT, 6J7GTX,

Original	Base Diagram	Preferred Substitution	Good Substitution
			6J7MG
7025	9A	7025A, 5751, 6057, 6681	12AX7, 12AX7A, 12AX7WA, 12AX7A/ECC83
7025A	9A	7025, 5751, 6057, 6681	12AX7, 12AX7A, 12AX7WA, 12AX7A/ECC83
7027	8HY	7027A	
7027A	8HY	7027	
7034		4X150A	
7036	7CH	5915	
7044	9H		
7054	9GK	8077	12BY7, 12BY7A
7055	6BT		
7056	7CM		
7057	9AJ		
7058	9EP		
7059	9AE		
7060	9DA	12CT8	
7061	9EU		
7062	9A	5965, 5965A	
7079	8DG	6111, 8526	6BF7
7083	5702	6540, 6872, 5702, 6245	
7105	6BD	6080, 6080WA, 6AS7, 6AS7G, 6AS7GA	
7137	7BQ	7245, 8522	
7167	7EW		
7184	7AC		6V6, 6V6G, 6V6GT, 6V6GTA, 6V6GTX, 6V6GTY, 6V6GX, 6V6Y
7189	9CV		6BQ5, 6BQ5/EL84

Original	Base Diagram	Preferred Substitution	Good Substitution
7199	9JT		
7203		4CX250B	
7211			
7212	7CK	6146, 8298	
7236	8BD	5998	
7244	7BF	7244A, 5964, 6045, 6101	6J6, 6J6A, 6J6WA, 6J6WB
7244A	7BF	7244, 5964, 6045, 6101	6J6, 6J6A, 6J6WB, 6J6WB
7245	7BQ	7245A, 8532	6J4, 6J4WA
7245A	7BQ	7245, 8532	6J4, 6J4WA
7247	9A	6189, 7730	12DW7
7258	9DA		
7289		3CX100A5	
7308	9AJ		
7318	9A	5814, 5814A, 6955, 7730, 6680, 12AU7, 12AU7WA, 12AU7A/ECC82	
7320	9CV	6BQ5	
7355	8KN		
7357	7CK		6159
7360	9KS		
7370	9H		5687
7408	7AC	6V6, 6V6G, 6V6GT, 6V6GTA, 6V6GTX, 6V6GTY, 6V6GX, 6V6Y	
7457			
7480			
7494	9A		12AX7, 12AX7A, 12AX7A/ECC83, 12AX7WA, 6681
7499	6F	6132	6CH6
7525		B166, 4-1000A	

150

Original	Base Diagram	Preferred Substitution	Good Substitution
7527		4-400B	
7543	7BK	6136, 8425	6AU6, 6AU6A
7551	9LK		
7558	9LK		
7576	8KM	8185	
7580		4CX250R	
7581	7AC	7581A, 6L6, 6L6A, 6L6G, 6L6GA, 6L6GB, 6L6GC, 6L6GT, 6L6GX, 6L6Y, 6L6W, 6L6WA, 6L6WB, 6L6WGT	
7581A	7AC	7581, 6L6, 6L6A, 6L6G, 6L6GA, 6L6GB, 6L6GC, 6L6GT, 6L6GX, 6L6Y, 6L6W, 6L6WA, 6L6WB, 6L6WGT	
7586	12AQ	8382	
7587	12AS	8380	
7591	8KQ	7591A	
7591A	8KQ	7591	
7609			
7645	9HL	6939	
7695	9PX		
7698			
7700	6F	6C6	
7701	9MS		7551
7717	7EW	6CY5, 8113	
7721	9EQ		
7722	9EQ		7721
7724	9KR	12GT8	
7728	9A	6201, 6679	12AT7, 12AT7/ECC81,

(continued)

Original	Base Diagram	Preferred Substitution	Good Substitution
			12AT7WA, 12AT7WB
7729	9A	6681, 7025, 12AX7, 12AX7A, 12AX7A/ECC83, 12AX7WA	
7730	9A	6680, 7247, 7318, 6189, 5814, 5814A, 12AU7A	12AU7
7731	9AE	6678, 7059, 6U8, 6U8A, 6KD8	
7732	7CM	6676	6CB6, 6CB6A, 6CF6
7733	9BF	5814, 5814A, 5963	12BY7, 12BY7A, 12BV7, 12DQ7
7734	9LC	6GE8	
7752	7CM	6AS6	
7759	8DG	7887	
7760	8DG	8103	
7788	9NK		
7803	9AJ		6FW8
7815		3CPN10A5	
7815AL			
7815R		3PX100A5	
7815RAL			
7815X			
7815XAL			
7843			
7855			
7855AL			
7855K			
7855KAL			
7868	9RW		
7887	8DG	7759	
7895	12AQ	6CW4, 6DS4	

Original	Base Diagram	Preferred Substitution	Good Substitution
7898	9EP		
7905	9PB		
8016	3C	1B3, 1B3GT, 1G3, 1G3GT, 1G3GTA	
8020		100R	
8032	7CK	6883	
8056	12AQ	8456	
8058	12CT		
8064	8DL	5906	
8070	8LD	8319	
8077	9GK	7054	
8103	8DG	7760	
8106	9PL		
8113	7EW	7717	
8136	7CM		6CB6, 6CB6A, 6CF6, 6DK6
8158		3CX10000A1	
8159		3CX10000A3	
8160		3CX10000A7	
8161		3CX2500A3	
8162		3CX3000F7	
8163		3-400Z	
8164		3-1000Z	
8165		4-65A	
8166		4-1000A	
8167		4CX300A	
8168		4CX1000A	
8169		4CX3000A	
8170		4CX5000A	
8170W		4CX5000R	
8171		4CX10000D	
8172		4X150G	
8187		4PR65A	
8188		4PR400A	
8189		4PR1000A	

Original	Base Diagram	Preferred Substitution	Good Substitution
8196	7CM	5754	6AS6
8203	12AQ		
8204	7BN	5727	2D21, 2D21, 2D21W
8233	9PZ		
8238		3CX3000A1	
8239		3CX3000F1	
8240		3CW5000A1	
8241		3CW5000F1	
8242		3CW5000A3	
8243		3CW5000F3	
8244		4CW2000A	
8245		4CX250K	
8246		4CX250M	
8247		4PR125A	
8248		4PR250C	
8249		4W300B	
8250		3X100F5	
8251		3CX2500F3	
8252		4PR60B	
8252W		4PR60C	
8255	9NY	6DL4, 6DL4/EC88	
8281		4CX15000A	
8283		3CX1000A7	
8295A			
8298	7CK	6146, 7212	
8321		4CX350A	
8322		4CX350F	
8349		4CX35000C	
8351		4CV100000C	
8352		4CX1000K	
8380	12AS	7587	
8382	12AQ	7586	
8393	12AQ		

154

Original	Base Diagram	Preferred Substitution	Good Substitution
8403			
8414	8DC	5908	
8417	8LY		
8425	7BK		6AU6, 6AU6A
8426	7BK		12AU6
8432			
8438		4-400A	
8441	12AQ	7895	
8445	9AE		6LN8, 6LN8/LCF80
8456	12AQ	8056	
8489	9DA		6AN8, 6AN8A
8522	8DC	8524, 5636	
8524	8DC	8522, 5625	
8525	8DG	6021, 6947	
8526	8DG	6111	
8527	8DK	5897	
8528	8DE	5902, 6224	
8529	8DE	5899, 8530	
8530	8DE	5840, 5900, 5901, 5906, 6049	
8532	7BQ	7245, 8532W	6J4, 6J4WA
8532W	7BQ	8532, 7245	6J4, 6J4WA
8533W			
8552	7CK	6883	
8556	12AQ	8056	
8560A			
8560AS			
8561		4CX300Y	
8576		264	
8590		4CPX250K	
8621		4CX250FG	
8627	12CT	8627A	
8627A	12CT	8627	

Original	Base Diagram	Preferred Substitution	Good Substitution
8628	12AQ		
8660		4CX1500B	
8661		4CW10000A	
8745			
8755			
8755A			
8756			
8757			
8808	8808		
8809		4CX600J	
8847			
8847A			
8873			
8874			
8875			
8876			
8877		3CX1500A7	
8892			
8893			
8904		4CX350FJ	
8906			
8906AL			
8906BAL			
8906X			
8906XAL			
8907			
8909		4CX5000J	
8910		4CX15000J	
8911			
8912			
8921		4CX600JA	
8930			
8933			
8938			

156

Original	Base Diagram	Preferred Substitution	Good Substitution
8940			
8941			
8942			
8944			
8950	8950		
8954			
8957		4CX250BC	
8959			
8960			
8961		3CX400U7	
8962			
8963			
8964			
8965			
9001	7BD		
9002	7BS		HA2, E1C, 5731
9003			E2F
9005	5BG		
9006	6BH		
18042			6086
18043			6689, E83F
18045			H81L
18046			1L861
18048			C3m
A61	9CB	17Z3. 17Z3/PY81	
A677	6F	6C6	
A863	7R		6J7, 6J7G, 6J7GT, 6J7GTX, 6J7MG
A1834	8BD	6080, 6080W	6AS7, 6AS7G, 6AS7GA, ECC230, 6520
A2252	5675		5675
A2275			5831

Original	Base Diagram	Preferred Substitution	Good Substitution
A2317			5893
A2351	5675		5876
A2352	5675		5675
A2354			5794
A2521	9BX		6CR4
A2599			6CT4
A2647			436A
A2900	9A	6201, 12AT7, 12AT7/ECC81	
A4051	5AW	807	
AA91E	6BT	5726, 6AL5, 6AL5W, 6663	
ABC91	7AC	12A6	
AC044			PP3/250, PX4
AC55		8170, 4CX5000A	
AD	4G	6Z3	
AFX212	5AY	6D4	
AG	4C	83	
AG866A	4P	866A	
AG2509	5BO	5651	
AG5210	5BO	0B2, 0B2WA, 6074	
AG5211	5BO	0A2, 0A2WA, 6073, 6626	
AH201	4P	866A	
AH216	4AT		872A
ARS25	5AW	807	
ASG512	7BN	2D21, 5727	
ASG5121	7BN	2D21, 5727	
ASG5023	3G	3C23	
ATS25	5AW	807	5B250A, QV05-25, 4Y25, QE06-50, P17A
AX224	4P	3B28	

Original	Base Diagram	Preferred Substitution	Good Substitution
AY365		8165, 4-65A	
AZ31			U143
B36	8BD	12SN7, 12SN7GT, 12SN7GTA	13D2
B63	7B	6A6	
B65	8BD	6SN7, 6SN7A, 6SN7GTA, 6SN7GTB, 6SN7GTY, 6SN7L	6CC10, ECC32, 6H8C
B109		28AQ8	10L14, UCC85
B139		7AN7	
B152	9A	12AT7, 12AT7/ECC81, 12AT7WA, 12AT7WB	
B309	9A	12AT7, 12AT7/ECC81, 12AT7WA, 12AT7WB	
B319	9FC	6CW7	6L16, ECC84
B329	9A	12AU7, 12AU7A, 12AU7WA, 12AU7A/ECC82	6CC40, ECC82
B339	9A	12AX7, 12AX7A, 12AX7WA, 12AX7A/ECC83, 5751, 6681	6C13, ECC83
B349		7EX7	30L15, PCC805
B719	9AJ	6AQ8, 6AQ8/ECC85	6L12, ECC85, 6CC43
B729		6GA8	ECC804
B739	9A	12AT7, 12AT7WA, 12AT7WB, 12AT7/ECC81	ECC81
B749	9A	12AU7, 12AU7A, 12AU7WA, 12AU7A/ECC82	ECC82

Original	Base Diagram	Preferred Substitution	Good Substitution
B759	9A	12AX7, 12AX7A, 12AX7WA, 12AX7A/ECC83	ECC83
B1109		3C24	
B1135		100TH	
BA2	6BS	2050, 2050A	
BF61		6CK5	EL41, N150
BF62			N151, EL42
BF451			45A5, N142, UL41
BPM04	7BZ	6AQ5, 6AQ5A, 6HG5	
BVA264		6AG6	
BVA265		6AG6	
BW194		6696A	
C3m			18048
C112		6156	
C143			813
C180			832
C610			757
C866	4P	866A	
C1108		4-125A	
C1112		5D22, 4-250A	
C1136		8438, 4-400A	
CC81E	9A	12AT7WA, 12AT7WB, 6201	12AT7, 12AT7ECC81, 6679
CC86E	9DE	6GM8	ECC86
CCa	9AJ	6922, 6922/E88CC	
CF61			ECH41
CF141			UCH41
CK502AX			DL71
CK505A			DF70
CK1003	4R	0Z4, 0Z4A, 0Z4G	

Original	Base Diagram	Preferred Substitution	Good Substitution
CR27	4P	866A	
CSF80	9AE	4BL8, 4BL8/XCF80	
CV216	4AJ	0D3, 0D3A	
CV427		8252W, 4PR60C	
CV449	5BO	5651	
CV475	8DE	5899	CV477
CV477	8DE	5899	CV475
CV618	4C	83	
CV688		35T	
CV686	4AJ	0C3, 0C3A	
CV752	4V	0A4G	CV1992
CV789		3C24	
CV807	7BB	3A4	CV2390
CV824		4-125A	
CV998		2000T	
CV1102		5D22, 4-250A	
CV1350		5867A	
CV1758	6AR	1L4	
CV1905		8165, 4-65A	
CV1832	5BO	0A2, 0A2WA, 6073, 6626	
CV1833	5BO	0B2, 0B2WA, 6074	
CV1834	8BD	6AS7, 6AS7G, 6AS7GA, 6080 6080WA	
CV1992	4V	0A4G	CV752
CV2129	9K	5763	
CV2130		6155	
CV2131		6156	
CV2159		7034, 4X150A	
CV2240	7CY		3B4WA
CV2241	5642	5642	
CV2390	7BB	3A4	CV807

Original	Base Diagram	Preferred Substitution	Good Substitution
CV2416		8252W, 4PR60C	
CV2466	9HL	6939	
CV2487		7203, 4CX250B	
CV2492	9AJ	6922, 6922/E88CC	
CV2516		7289	
CV2519		7304, 4X150A	
CV242222	7CM	6AS6, 5725	
CV2552		100TH	
CV2572		450TH	
CV2573	5BO	5651, 5651A, 5651WA	CV048, CV5186
CV2589		250TH	
CV2611		304TH	
CV2642	9V	5842, 417A	CV3789
CV2662	8DE	5639	
CV2711		1500T	
CV2742	6AR	1L4	CV2795
CV2752		8252W, 4PR60C	
CV2795	6AR	1L4	CV2742
CV2876	7BN	5727, 2D21, 2D21W	
CV2963		4-125A	
CV2964		5D22, 4-250A	
CV2967		8020, 100R	
CV2984	8BD	6080, 6080WA	6AS7, 6AS7G, 6AS7GA
CV3508	9A	6201, 12AT7, 12AT7WA, 12AT7WB, 12AT7/ECC81, 6679	
CV3512	7BN	5696	
CV3789	9V	5842, 417A	CV642
CV3798	4AJ	0A3, 0A3A	KD21, QS1205

Original	Base Diagram	Preferred Substitution	Good Substitution
CV3879		8438. 4-400A	
CV3880		8166. 4-1000A	
CV3893		8172. 4X150G	
CV3928	8DC	5836	5840, 5840A. 5840W
CV3929	8DE	5840. 5840A. 5840W	
CV3930	8DK	5718	
CV3986	8DG	6021	
CV3991		7609	
CV4008	8DK	5719	
CV4009	7KB	5749	6BA6. 6BA6/EF93
CV4011	7CM	5725	6AS6
CV4016	9A	5814. 5814A	
CV4017	9A	5751	6681, 12AX7. 12AX7A. 12AX7WA. 12AX7A/ECC83
CV4018	7BN	5727, 2D21. 2D21W	M8204. E91N
CV4020	5BO	0A2. 0A2WA	HD51, QS1207. AG5211
CV4023	7BK	6AU6. 6AU6A	
CV4024	9A	12AT7, 12AT7WA. 12AT7WB. 12AT7/ECC81. 6679	
CV4025	6BT	5726	6AL5. 6663
CV4028	5BO 6074	0B2. 0B2WA.	CV1833, HD52. M8224
CV4029	8DE	5902	
CV4031	7BF	6101	6J6, 6J6A. 6J6WA. 6J6WB
CV4039	9K	5763	
CV4048	5BO	5651. 5651A.	CV2573

Original	Base Diagram	Preferred Substitution	Good Substitution
		5651WA	
CV4100	5BO	0A2, 0A2WA, 6626, 6073	AG5211, CV4020, HD21
CV4101	5BO	0B2, 0B2WA, 6074	AG5210
CV5122	4CK	5823	
CV5176		2-01C	
CV518686	5BO	5651, 5651A, 5651WA	CV4048, CV2473
CV5212	9A	6201	12AT7, 12AT7WA, 12AT7WB, 6679
CV5430		7289	
CV5959		7527, 4-400B	
CV6122		8165, 4-65A	
CV6131		8232W, 4PR60C	
CV6137		7203, 4CX250B	
CV6184		8171, 4CX10000D	
CV8295		8170, 4CX5000A	
CV8698		8321, 4CX350A	
CV11106		5CX1500A	
CV11107		8349, 4CX35000C	
D1C	5BD	957	
D2C	5BD	958	
D2M9	6BT	5726	6AL5, 6663
D27	6BT	5726	6AL5, 6663
D61			2B35, 6D1, EA50
D63	7Q		6H6, 6H6G, 6H6GT, 6H6MG
D77	6BT	5726, 6AL5, 6663	E91AA, EAA91, EB91, 6D2, 6B32
D121			UAF41
D152	6BT	5726, 6AL5, 6663	E91AA, EAA91, EB91, 6D2, 6B32

164

Original	Base Diagram	Preferred Substitution	Good Substitution
D717	6BT	5726, 6AL5, 6663	
DA90	5AP	1A3	1D13
DAC21	6AU	1S5	
DAC32	5Z	1H5	HD14
DAF90	5AP	1A3	
DAF91	6AU	1S5	1FD9, ZD17
DAF92	6BW	1U5	1DN5
DAF96	1AG5	1AJ5	1FD1, ZD25
DAF97			1AN5
DC70			DC761
DC80			1E3
DC761			DC70
DCC90	7BC	3A5	
DCF60	1V6	1V6	
DCG4	4P	866A	
DD6	6BT	5726, DD6G	6AL5, 6663
DD6G	6BT	5726, DD6	6AL5, 6663
DD7			6AM5
DD77	6BT	5726	6AL5
DDR7			6AM5
DET18		35T	
DET77			810
DET22			5861, EC55, TD03-10
DF26	6AU	1S5	
DF33	5Y	1N5	Z14
DF60	1AD4	5678	
DF62	1AD4	1AD4	
DF67			5911, DF167
DF70			CK505AX
DF91	6AR	1T4	1F3, W17
DF92	6AR	1L4	1F2
DF96			1F1, W25, 1AJ4

Original	Base Diagram	Preferred Substitution	Good Substitution
DF97			1AN5
DF167			5911
DF650			6419
DF652	1AD4	1AD4	
DF654	1AD4	5678	
DF668	1AD4	1AD4	
DF669	1AD4	5678	
DF703	5886	5886	
DF904	6AR	1U4	
DH63	7V	6Q7, 6Q7G, 6Q7GT, 6Q7MG	6B6
DH73			6Q6, 6V7
DH74	7V	12Q7	
DH76	7V	12Q7	DL74M
DH77	7BT	6AT6, 6BT6	EBC90
DH81	8W	7B6	
DH109			28AK8, 10LD12, UABC80
DH118			14L7, 10LD3, UBC41
DH119			15BD7A, 10LD13, 14G6, UBC81
DH142			14L7, 10LD3, UBC41
DH147			EBC33
DH149	8W	7C6	
DH150			6CV7, 6LD3, EBC41
DH718			6CV7, 6LD3, EBC41
DH719	9E	6AK8, 6AK8/EABC80	6LD12, 6DT31, EABC80
DH817			6CV7
DK32	7Z	1A7, 1A7GT, 1B7, 1B7GT,	X14
DK91	7AT	1R5	1C1, X17

166

Original	Base Diagram	Preferred Substitution	Good Substitution
DK92			1C2, X20, 1AC6
DK96			1C3, X25, 1AB6, 1H35
DK97			1AB6
DL29	6BB	3LF4	3D6
DL31	6X	1A5	1T5
DL33	7AQ	3B5, 3Q5, 3Q5G, 3Q5GT	
DL35	6X	1C5	N14
DL36	6AF	1Q5	
DL37	7AC	6L6, 6L6A, 6L6G, 6L6GA, 6L6GB, 6L6GC, 6L6GT, 6L6GX, 6L6Y, 6L6W, 6L6WA, 6L6WB, 6L6WGT	
DL63	7V	6R7	
DL67			5913, DL167
DL70			6373
DL71			CK502AX
DL72			CK506AX
DL74M	7V	12Q7GT	DH76
DL82	8W	7B6	
DL91	7AV	1S4	
DL92	7BA	3S4	
DL93	7BB	3A4	
DL94	6BX	3V4	1P11, N19
DL95	7BA	3Q4	N18
DL96	6BX	3E5	1P1, N25
DL98	7CY	3B4WA	
DL145			10LD11
DL167			5913
DL620	2E31	5672	DL652
DL652	2E31	5672	
DL700	9BV	6197	

Original	Base Diagram	Preferred Substitution	Good Substitution
DL761	6CL	6147, 5851	
DL012	9E	6T8, 6T8A	
DM70	8GV	1M3	1M1, Y25
DM71	8GV	1N3	
DM160	6977	6977	
DN143			EBL21, EBL71
D024			PP4/500, PX25
D026			PP4/500, PX25
DP61	7BD	6AK5, 6AK5/EF95	EF95
DX144			EC56
DX145	8108	8108	EC57, EC157
DY30	3C	1B3, 1B3GT, 1G3, 1G3GT, 1G3GTA	
DY51	9RG	1BC2	
DY70	5642	5642	
DY80	9Y	1X2, 1X2A, 1X2B, 1X2C	R19, 1R6
DY86	9DT	1S2, 1S2A, 1AX2, 1S2A/DY87	DY87
DY87	9DT	1S2, 1S2A, 1S2A/DY87	
DY87			1BQ2
E1C	5BC	955	HA2
E1F	5BB	954	ZA2
E1T			6370
E2F	5BB	956	
E20C			437, 3A/167M
E55L	9PZ	8233	
E80CC	9A	6085	ECC87
E80CF	9AE	6AX8	6C16
E80F	9BJ	6084	
E80L			6227
E80T			6218

Original	Base Diagram	Preferred Substitution	Good Substitution
E81CC	9A	12AT7, 12AT7WA, 12AT7WB, 12AT7/ECC81, 6201, 6679	B162, ECC81
E81H			EH81
E81L	9AU	6686	EL861
E82CC	9A	5814, 5814A, 6189, 6680, 12AU7, 12AU7A, 12AU7A/ECC82	B329, ECC82, 6CC40
E82M			5624
E83CC	9A	6681, 12AX7, 12AX7WA, 12AX7A, 12AX7A/ECC83	
E83F			6689
E84L	9CV	7189	
E86C			EC806S
E88C			8562
E88CC	9AJ	6922	
E89F			6DG7
E90C	9BF		5920
E90CC	9BF	5920	ECC960
E90F	7CM	6BH6	
E90Z	5BS	6063, 6X4, 6X4W	EZ900
E91AA	6BT	5626, 6AL5, 6663	6D2, D77, 6B32
E91H	7CH	5915, 5915A	EH900/S
E91N	7BN	5727, 2D21, 2D21W	CV4018, M8204
E92CC			E99CC
E95F	7BD	5654, 6AK5, 6AK5/EF95	EF95, 6F32V
E99CC			E92CC
E99F	7CM	6662, 6BJ6	
E130L			7534

Original	Base Diagram	Preferred Substitution	Good Substitution
E180CC			5962
E180F	9EQ	6688, 6688A	EF861, 5A/170K
E180L			7534
E181CC			7118
E182CC			7119
E182F	9X	5847, 404, 404A	
E186F	9MZ	7737	
E188CC	9DE	6962, 7308	
E235L	8KB	7751	
E236L			6CM5
E250A		6156	
E280F	9EQ	7722	
E283CC	9A	12DT7	6L13, B339
E288CC	9AJ	8223	
E810F	9NK	7788	
E900		250TH	
E902	5BS	6X4, 6X4W	
E1485	7BB	3A4	DL93
E1955	7BN	2D21, 2D21W, 5727	EN91, WT210-0001
E2157	9A	12AT7, 12AT7WA, 12AT7WB, 12AT7/ECC81	
E2163	9A	12AU7, 12AU7A, 12AU7WA, 12AU7A/ECC882	
E2164	9A	12AX7, 12AX7A, 12AX7WA, 12AX7A/ECC83, 6681	
E3033		8187, 4CX10000D	
EA41			6CT7
EA50			2B35, D61, 60¹

Original	Base Diagram	Preferred Substitution	Good Substitution
EA52			6923
EA76			6489
EA766	5704	5704	
EAA91	6BT	5726, 6AL5, 6663	D152
EAA901	6BT	5726, 6AL5, 6663	D152
EAA901S	6BT	5726, 6AL5, 6663, 6058	M8079
EABC80	9E	6AK8, 6AK8/EABE80	6LD12, DH719, 6DT31
EAC91			M8097
EAF42			6CT7
EAM86			6GX8
EB34	7Q	6H6, 6H6G, 6H6GT, 6H6MG	D63
EB91	6BT	6AL5, 6097	
EBC3			6BD7A
EBC33			DH147
EBC41			6CV7, DH150, 6LD3
EBC80		6BD7	
EBC81		6BD7	6LD13
EBC90	7BT	6AT6	DH77
EBC91	7BT	6AQ6, 6AV6	
EBF32	8E	6B8	
EBF41			6CJ5
EBF80		6N8	WD709
EBF81		6AD8	
EBF83			6DR8
EBF85	9HE	6DC8, 6DC8/EBF89	EBF89
EBF89	9HE	6DC8, 6DC8/EBF89	
EBL21			DN143, EBL71

Original	Base Diagram	Preferred Substitution	Good Substitution
EBL71			DN143, EBL21
EC20	8DK	6AK4	6K, 5718, 6778, EC71, SN979
EC22			6R4
EC32	8FU	6BD4, 6BD4A	
EC51			5861
EC55			5861, DET22, TD03-10
EC56			DX144
EC57	8108	8108	
EC70	8DK	5718	
EC71	8DK	5718	
EC80			6Q4
EC81			6R4
EC84	9BX	6AJ4, 6AM4	
EC86			6CM4
EC88	9NY	6DL4, 6DL4/EC88	
EC90	6BG	6C4	
EC90			6AQ4, 6L34
EC92	5CE	6AB4, 6664	
EC93			6BS4, EC903
EC94	7DK	6AF4, 6AF4A	6AN4, 6T1, 6T2, 6DZ4
EC95	7FP	6ER5	
EC97	7FP	6FY5	
EC98	7BQ	8532	
EC157	8108	8108	
EC158			8436
EC560	2C39	2C39	
EC562	2C40	2C39	
EC760			5718
EC806S			E86C
EC900	7GM	6HA5, 6HM5, 6HK5	

172

Original	Base Diagram	Preferred Substitution	Good Substitution
EC903			6BS4
EC1000	8LW	8254	
EC8008S	9NY	8255	
EC8010			8556
ECC32	8BD	6SN7, 6SN7A, 6SN7GTA, 6SN7GTB, 6SN7GTY, 6SN7L	6H8C, B65
ECC35	8BD	6SL7, 6SL7A, 6SL7GT, 6SL7GTY, 6SL7	
ECC40			AA61
ECC70	8DG	6021	6BF7, 6BG7
ECC81	9A	6060, 12AT7, 12AT7WA, 12AT7WB, 12AT7/ECC81	B309
ECC82	9A	12AU7, 12AU7A, 12AU7WA, 12AU7A/ECC82, 6189	6CC40, B329
ECC83	9A	12AX7, 12AX7A, 12AX7WA, 12AX7A/ECC83, 6681	B339, 6L13
ECC84	9A	6CW7	B319, 6L16
ECC85	9AJ	6AQ8, 6AQ8/ECC85	B719, 6L12
ECC86	9DE	6GM8	
ECC88	9DE	6DJ8, 6DJ8/ECC88, 6ES8, 6ES8/ECC189	
ECC89	9DD	6FC7	
ECC91	7BF	6J6, 6J6A, 6J6WA, 6J6WB, 6101	6CC31

Original	Base Diagram	Preferred Substitution	Good Substitution
ECC180	9AJ	6BQ7, 6BQ7A, 6BZ7, 6BS8	
ECC186	9A	12AU7, 12AU7A, 12AU7WA, 12AU7A/ECC82, 7316	ECC82
ECC189	9DE	6ES8, 6ES8/ECC189, 6DJ8, 6DJ8/ECC88	
ECC230	8BD	6080, 6080WA, 6AS7, 6AS7G, 6AS7GA	A1834
ECC801	9A	6201, 12AT7, 12AT7WA, 12AT7WB, 12AT7/ECC81, ECC801/S	A2900
ECC801/S	9A	ECC801, 6201, 12AT7, 12AT7WA, 12AT7WB, 12AT7/ECC81	A2900
ECC802	9A	6067, 6189, 6080, 12AU7, 12AU7A, 12AU7WA, 12AU7A/ECC82, ECC802/S	M8146
ECC802/S	9A	ECC802, 6067, 6080, 6189, 12AU7, 12AU7A, 12AU7WA, 12AU7A/ECC82	M8136
ECC803	9A	ECC803/S, 12AX7, 12AX7A, 12AX7WA, 12AX7A/ECC83	

174

Original	Base Diagram	Preferred Substitution	Good Substitution
ECC803/S	9A	ECC803, 12AX7, 12AX7A, 12AX7WA, 12AX7A/ECC83	
ECC804			6/30L2, 6GA8, B729
ECC805			6L15
ECC807			1307
ECC808			6KX8
ECC813	9DE	6GM8	
ECC863			12DT7
ECC865	9FC	6CW7	
ECC900	7GM	6HA5, 6HM5, 6HK5	
ECC960			5920
ECF80	9AE	6BL8, 6BL8/ECF80	6C16, EC80F
ECF82	9AE	6U8, 6U8A, 6U8A/6KD8, 6KD8	6TP1
ECF86	9MP	6HG8, 6HG8/ECF86	
ECF200	10K	6X9, 6X9/ECF200	
ECF201	10K	6U9, 6U9/ECF201	
ECF202			6AJ9
ECF801	9QA	6GJ7, 6GJ7/ECF201	6GX7
ECF802	9DC	6JW8, 6JW8/ECF802	
ECF805	9KN	6GV7	6C18
ECH21			X143, ECH71

Original	Base Diagram	Preferred Substitution	Good Substitution
ECH33			6E8, X65
ECH35	8K	6K8, 6K8G, 6K8GT, 6K8GTX	X61M, 6C31
ECH41			CF61
ECH42		6C9	6CU7, ECH113, Z150
ECH43	8BL	7J7, C610	Z150
ECH71			X143
ECH80			6AN7
ECH81	6E	6AJ8, 6AJ8/ECH81	6C12, X719, 20D4
ECH82			6E8
ECH83			6DS8
ECH84			6JX8
ECH113	8BL	6C9	6CU7, Z150
ECH200			6V9
ECL80	6AB8	6AB8	LN152
ECL82	9EX	6BM8, 6BM8/ECL82	6PL12
ECL84	9HX	6DX8, 6DX8/ECL84	6DQ8
ECL85	9LY	6GV8, 6GV8/ECL85	
ECL86	9LZ	6GW8, 6GW8/ECL86	
ECL100	9AJ	6BQ7, 6BQ7A, 6BZ7, 6BS8	
ECL180	9AJ	6BQ7, 6BQ7A, 6BZ7, 6BS8	
ECL821			6CH6
ECLL800			6KH8
ED2	6BT	6AL5	
ED500			6ED4
EE80			6F28
EF2			6DA6

Original	Base Diagram	Preferred Substitution	Good Substitution
EF5			6DA6
EF13			6DA6
EF22	8V	7G7	
EF36	7R	6J7, 6J7G, 6J7GT, 6J7GTX, 6J7MG	
EF37	7R	6J7, 6J7G, 6J7GT, 6J7GTX, 6J7MG	
EF39	7S	6K6, 6K6G, 6K6GT, 6K6MG	9D4, W147
EF41			6CJ5, 6F15, W150
EF42		6F13, Z150	
EF70			6487
EF71	8DL	5899	
EF72	8DE	5840, 5840W	
EF80	9AQ	6BW7	8D6, Z152, 6F41
EF81			6BH5
EF82			6CH6, 7D10, EL821
EF83			6BK8
EF85	9AQ	6BY7	6F19, W719, EF805S
EF86	9CQ	6267	6CF8, Z729, 8D8, 6F22
EF87	9CQ		6CF8, 6267
EF89			6DA6
EF89F			6DG7
EF91	7BD	6AM6, 6AM6/EF91	6AM6
EF92	7BD	6CQ6	W77, 9D6, 6F21
EF93	7BK	6BA6, 6BA6/EF93, 5749, 6660	W727, 6F31

Original	Base Diagram	Preferred Substitution	Good Substitution
EF94	7BK	6AU6, 6AU6A, 6136	6P2
EF95	7BD	6AK5, 6AK5/EF95, 5654	6F32V, 403, 403A, 403B, 5591
EF96	7BD	6AG5	
EF97	7BK	6ES6	
EF98	7BK	6ET6	
EF183	9AQ	6EH7, 6EH7/EF183	6F29
EF184	9AQ	6EJ7, 6EJ7/EF184	6F30
EF190	7CM	6CB6, 6CB6A, 6BZ6	
EF730	8DC	5636	
EF731	8DL	5899	
EF732	8DE	5840	
EF734	8DC	6205	
EF762	8DL	5901	
EF800	9AQ	6BX6	Z719, 6P6
EF804			6F40
EF811	9AQ	6EH7, 6EH7/EF183	6F25
EF814	9AQ	6EJ7, 6EJ7/EF184	6EL7, 6F23
EF814	9AQ		6F24
EF860	9AQ	6BX6	EF800, Z719, 6P6
EF861	9EQ	6688	5A/170K
EF866			6CF8
EF903			6BS4
EF905	7BD	6AK5, 6AK5/EF95	6F32V, 403, 403A, 403B
EF8055	9AQ	6BY7	W719, 6F19

Original	Base Diagram	Preferred Substitution	Good Substitution
EF8065			6BK8, Z729, 8D8, 6F22
EFL200	10L	6Y9, 6Y9/EFL200	6W9
EH90	7CH	6CS6, 6BY6, 6BX6	
EH900S	7CH	5915, 5915A	
EH960			6687
EK90	7CH	6BE6, 5750	X77, X727
EL22	6AE	7B5	
EL31			5P29, 6CN6, EL38
EL32			1637
EL33			6P25, KT51, 6M6, 6PX6, 6AG6
EL34	8ET	6CA7, 6CA7/EL34	12E13, 7D11, KT88
EL35	7AC	6L6, 6L6A, 6L6G, 6L6GA, 6L6GB, 6L6GC, 6L6GT, 6L6GX, 6L6Y, 6L6W, 6L6WA, 6L6WB, 6L6WGT	N66
EL36	6AM	6GW6, 6DQ6, 6DQ6A, 6DQ6B	6CM5, E23LL
EL37	7AC	5881, 6L6, 6L6A, 6L6G, 6L6GA, 6L6GB, 6L6GC, 6L6GT, 6L6GX, 6L6Y, 6L6W, 6L6WA, 6L6WB, 6L6WGT	
EL38			EL31, 6CN6, 5P29
EL39	7AC		KT66, 6CN5

Original	Base Diagram	Preferred Substitution	Good Substitution
EL41			6CK5, N150, BF61
EL42			N151, BF62
EL70			7001, M8167
EL37	5640	5640	
EL80		6M5	
EL81			6CJ6, EL820
EL82	9CV		6DY5
EL83			6CK6, EL803/S
EL84	9CV	6BQ5, 6BQ5/EL84	N709, 6P15, 7189
EL85			6BN5, N155
EL86	9CV	6CW5, 6CW5/EL86	
EL90	7BZ	6AQ5, 6AQ5A	N727, 6L31
EL91			6AM5, 6P17, 7D9, N77
EL95	7DQ	6DL5, 6DL5/EL95	
EL136			6FV5
EL180	9BF	12BY7, 12BY7A, 12BV7, 12DQ7	
EL300			6FN5
EL500	9NH	6GB5, 6GB5/EL500	
EL503			8278
EL504	9NH	6GB5, 6GB5/EL500	
EL505	9RJ	6KG6, 6KG6A, 6KG6A/EL509	EL509
EL508			6KW6
EL509	9RJ	6KG6, 6KG6A, 6KG6A/EL509	EL505
EL802			6LD6

Original	Base Diagram	Preferred Substitution	Good Substitution
EL803/S			6CK6
EL820	9AS		6CJ6
EL821	6F	6132	6CH6
EL861	9AU	6686	
EL862	9EQ	7721	D3a, 5A/85K
ELF86	9MP	6HG8, 6HG8/ECF86	ECF86
ELL80	9NJ	6HU8, 6HU8/ELL80	
EM34			6AF7, 6CD7, 6M1
EM35	6R	6U5, 6G5	Y61, 6M2
EM81	9DB	6DA5	
EM84	9GA	6FG6, 6FG6/EM84	
EM84a	9GA	6HU6, 6HU6/EM87	
EM85			6DU6
EM87	9GA	6HU6, 6HU6/EM87	
EM180			6BR5, 6M40
EM840	9GA	6FG6, 6FG6/EM84	
EN32	6BS	2050, 2050A	6574
EN70			5643
EN91	7BN	2D21, 3D21W, 5727	20A3
EN92	7BN	5696, 5696A	
EN93	5AY	6D4	
EQ80			6BE7
ESU866	4P	866A	
ET1000		250TH	
EY51			6X2, U43, R12
EY70	6CJ	5641	
EY80			6U3

Original	Base Diagram	Preferred Substitution	Good Substitution
EY81	9CB	6R3	
EY81F	9BD	6V3, 6B3	
EY82		6N3	
EY81F	9BD	6V3, 6B3	
EY82		6N3	
EY83	9CB	6AL3, 6AL3/EY88	6BR3
EY84		6374	
EY86			6S2
EY87			6S2A
EY88	9CB	6AL3, 6AL3/EY88	6BR3, EY83
EY89			6AV3
EY500	6EC4	6EC4, 6EC4A, 6EC4A/EY500	
EZ3	9M	6V4	
EZ4	9M	6CA4	
EZ11	9M	6V4	
EZ12			6GC6
EZ22	5AB	7Y4	U82
EZ35	6S	6X5, 6X5G, 6X5GT, 6X5L, 6X5MG, 6X5W, 6X5WGT	
EZ40			6BT4, UU9, U150
EZ80	9M	6V4	
EZ81	9DJ	6BW4	U709, UU12
EZ90	5BS	6X4, 6X4W	6Z31, U78
EZ91			6AV4, 6FX4
EZ900	5BS	6X4, 6X4W, 6063	
FA6	5676	5677	
G/50/4K	5BO	0A2, 0A2WA	
G75/2D	4AJ	0A3, 03AA	QS1205
G75/4K	5BO	0C2	

Original	Base Diagram	Preferred Substitution	Good Substitution
G77	6F	6C6	
G84			2Z2
G105/1D	4AJ	0C3, 0C3A	
G108/1K	5BO	0B2, 0B2WA	QS1208, 108C1
G150/3D	4AJ	0D3, 0D3A	QS150/40, 150C3
G150/4K	5BO	0A2, 0A2WA	QS1207, 150C2
G180/2M			QS150/45
GL546	7BN	5696, 5696A	
GU12	4P	866A	
GY86			3S2
GY501			3BH2
GZ30	5L	5Z4, 5Y3G, 5Y3GA, 5Y3GT	
GZ31	5T	5U4, 5U4G, 5U4GB	5Z10
GZ32	5L	5V4, 5V4G, 5V4GA	5AQ4
GZ34	5DA	5AR4, 5AR4/GZ34	5T4, U52, 5CG4
GZ37	5DA	5AR4, 5A54/GZ34	U54
GZ4			V51
H2-10	4AB	2X2	
H52	5T	5U4, 5U4G, 5U4GB	5AS4
H63	5M	6F5, 6F5G, 6F5GT, 6F5MG	
H81L			18045
H250	6F	6C6	
HAA91	6BT	12AL5	10D2, UB91
HABC80	9E	19T8	19C8
HBC90	7BT	12AT6	12BT6, 9TP9

Original	Base Diagram	Preferred Substitution	Good Substitution
HBC91	7BT	12AV6	12BC32
HCC85	9AJ	17EW8, 17EW8/HCC85	
HCH81			12AJ8, 12D8
HD14	5Z	1H5, 1H5GT	
HD30	7CY	3B4, 3B4WA	
HD51	5BO	0A2, 0A2WA, 6073, 6626	
HD52	5BO	0B2, 0B2WA, 6074	
HD93	9Y	1X2, 1X2A, 1X2B, 1X2C	
HD94	6AM	6BQ6, 6BQ6GT, 6BQ6GTB, 6CU6	
HD96	6AM	25BQ6, 25BQ6GA, 25BQ6GT, 25BQ6GTB, 25CU6	
HF61			6CJ5
HF62			EF42, 6F13, Z150
HF93	7BK	12BA6	12E4
HF94	7BK	12AU6	12ET1
HK90	7CH	12BE6	12H31
HL86	9CV	30CW5	
HL90			19AQ5
HL92	7CV	50C5	
HL94			30A5, 30C5
HM04	7CH	6BE6, 5750	
HP6			6AM6
HR1			6305, 2T/270K
HY51B			829B
HY90	5BQ	35W4	35R1
HY145	6AR	1U4	
HZ50			14Z3

Original	Base Diagram	Preferred Substitution	Good Substitution
HZ90	4G	12Z3	14Z3
IL861			18046
ITW101		3CW10000H3	
KD21	4AJ	0A3, 0A3A	
KD24	4AJ	0C3, 0C3A	
KD25	4AJ	0D3, 0D3A	
KF35	5Y	1E5	
KK32	7Z	1C7	
KT32	7AC	25L6, 25L6G, 25L6GT, 25W6, 25W6GT	
KT33	7S	25A6	
KT61			6AG6, 6P25, 6M6, 6PX6
KT63	7S	6F6, 6F6G, 6F6GT, 6F6MG	
KT66	7AC	6L6, 6L6A, 6L6G, 6L6GA, 6L6GB, 6L6GC, 6L6GT, 6L6GX, 6L6Y, 6L6W, 6L6WA, 6L6WB, 6L6WGT, 5881	
KT71	7AC	50L6, 50L6GT	50C6
KT77	8EP	6CA7, 6CA7/EL34	EL34
KT81			7C5
KT88	8EP	6CA7, 6CA7/EL34	
KTW61	7R	6S7, 6S7G	
KTW63	7R	6K7, 6K7G, 6K7GT, 6K7GTX, 6K7MG	
KTZ63	7R	6J7, 6J7G, 6J7GT, 6J7GTX, 6J7MG	

Original	Base Diagram	Preferred Substitution	Good Substitution
KY50			2L2, U47, U25
KY80			2J2, U49, R20, U26
L63	6Q	6C5, 6C5G, 6C5GT, 6J5, 6J5G, 6J5GT, 6J5GTX, 6J5GX, 6J5MG	
L77	6BG	6C4	EC90
LC97	7FP	3FY5, 3ER5, 3GK5	
LC900	7GM	3HA5, 3HM5	
LCF80	9AE	6LN8, 6LN8/LCF80	
LCF86	9MP	5HG8, 5HG8/LCF86	
LCF200	10K	5X9	
LCF201	10K	5U9, 5U9/LCF201	
LCF801	9QA	5GJ7, 5GJ7/LCF801	
LCF802	9DC	6LX8, 6LX8/LCF802	
LCH200			5V9
LCL82	9EX	11BM8	
LCL84	9HX	10DX8, 10DX8/LCL84	
LDR03			0RP12
LF183	9AQ	4EH7, 4EH7/LF183	
LF184	9AQ	4EJ7, 4EJ7/LF184	
LFL200	10L	11Y9, 11Y9/LFL200	

Original	Base Diagram	Preferred Substitution	Good Substitution
LL86	9CV	10CW5, 10CW5/LL86	
LL119	9EX	48A8	10PL12, UCL82
LL500	9NH	18GB5, 18GB5/LL500	
LL505	9RJ	27KG6	
LL521	9RJ	21KQ6	
LN119	9EX	50BM8, 50BM8/UCL82	
LN152	6AB8	6AB8	ECL80
LN309	9EX	16A8, 16A8/PCL82	PCL83
LN319			13GC8, 30PL1, PCL801
LN329			30PL14, PCL88
LN339			9GB8, 30FL1
LN369	9EX	16A8, 16A8/PCL82	PCL82, 30PL12
LY81			11R3
LY88	9CB	20AQ3, 20AQ3/LY88	
LY500			28EC4
LZ319	9DC	9A8, 9A8/PCF80	9C8, PCF80
LZ329	9DC	9A8, 9A8/PCF80	30C1, 8A8, PCF80
LZ339			9EN7, 30C15, PCF800
M8063	7BD	6AM6, 6AM6/EF91	
M8079	6BT	5726, 6AL5, 6058, 6663	
M8080	6BG	6100, 6C4, 6C4W, 6C4WA	
M8081	7BF	6101, 6J6, 6J6A, 6J6WA, 6J6WB	6M-HH3
M8082			6516

Original	Base Diagram	Preferred Substitution	Good Substitution
M8083			6024. 6F12. 5A/160K
M8091			6443
M8096	9K	5763. 6062	
M8097			EAC91
M8098	5BO	5651. 5651A. 5651WA	
M8100	7BD	5654. 6AK5. 6AK5/EF95. 6069	
M8101	7BK	6BA6. 6BA6/EF93. 5749. 6660	
M8136	9A	12AU7. 12AU7A. 12AU7WA. 12AU7A/ECC82. 6189	
M8137	9A	12AX7. 12AX7A. 12AX7WA. 12AX7A/ECC83	
M8138	5BS	6202	6X4. 6X4W. 6063
M8140			F/6064
M8144			F/6060
M8149			F/6067
M8161	7BK		6065
M8162	9A	6679. 12AT7. 12AT7WA. 12AT7WB. 12AT7/ECC81. 6201	
M8167			7001. 3L70
M8180	7BD	5654. 6AK5. 6AK5/EF95	
M8190	5783	5783	
M8195			6CF8
M8196	7CM	6AS6. 5725. 6AS6W	6F33

Original	Base Diagram	Preferred Substitution	Good Substitution
M8204	7BN	5727	E91N, CV4018
M8212	6BT	5726, 6663, 6AL5, 6097	EB91
M8214			F/6057
M8223	5BO	0A2, 0A2WA, 6627, 6073	
M8224	5BO	0B2, 0B2WA, 6074	
M8232	7BQ	8532, 6J4, 6J4WA	
M8237			F/5726
M8245	7BZ	6005, 6AQ5, 6AQ5A	
M8248	7BZ	8532	
ME1501	6BS	2050, 2050A	
MU14			6BT4
MV6-5	8R	6SA7, 6SA7G, 6SA7GT, 6SA7GTX, 6SA7GTY, 6SA7Y	
N14	6X	1C5, 1C5GT	
N15	7AQ	3B5, 3Q5, 3Q5G, 3Q5GT	
N16	7AQ	3B5, 3Q5, 3Q5G, 3Q5GT	
N17	7BA	3S4	DL92, 1P10
N18	7BA	3Q4	DL95
N19	6BX	3V4	1P11, DL94
N22LL			19FK6
N25	6BX	3E5	
N30EL	12GW	6LF6	
N47			6AM5
N63	7S	6K6, 6K6G, 6K6GT, 6K6MG	1621
N66	7AC	6L6, 6L6A, 6L6G, 6L6GA, 6L6GB,	

Original	Base Diagram	Preferred Substitution	Good Substitution
		6L6GC, 6L6GT, 6L6GX, 6L6Y, 6L6W, 6L6WA, 6L6WB, 6L6WGT	
N77			6AM5, 7D9, 6P17, EL91
N78	6CH	6BJ5	
N118			10P13
N119	9CV	45B5	10P10, 10P18, UL84
N142			45A5, BF451, UL41
N144	6CH		7D9, 6P17, EL91
N145			10P13
N147			6AG6, EL33, 6P25
N148	6AE	7B5	EL22
N150			6CK5, BF61, EL41
N151			BF62, EL42
N152	9AS	21A6	PL820, PL81
N153	9AR	15A6	PL83
N154	9BL	16A5	16L40, 30P16, PL82
N155			6BN5, EL85
N308	8GT	25E5, 25E5/PL36	25GF6, 30P4
N309	9AR	15A6	PL83
N329	9BL	16A5	16L40, 30P16, PL82
N339			21B6, 21L40, PL81
N359	9KS	21B6	21L40, PL81
N369	9EX	16A8, 16A8/PCL82	12BF5
N378	9CV	15CW5, 15CW5/PL84	15BQ5, 30P18, PL84

Original	Base Diagram	Preferred Substitution	Good Substitution
N379	9CV	15CW5, 15CW5/PL84	30P18, PL84
N389			30P19, PL302
N709	9CV	6BQ5, 6BQ5/EL84	6P15, EL84
N727	7BZ	6AQ5, 6AQ5A	EL90, 6L31
NE38	991	991	
0BC3	8Q	12SQ7, 12SQ7G, 12SQ7GT	
0F1	7R	6S7, 6S7G	
0F5	7R	12K7, 12K7GT	
OH4	8A	12A8, 12A8GT	
OM3	7Q	6H6, 6H6G, 6H6GT, 6H6MG	
OM6	7R	6K7, 6K7G, 6K7GT, 6K7GTX, 6K7MG	
ORP12			LDR03
OSW2190	8N	6AC7, 6AC7W	
OSW2192	8Y	6AG7, 6AG7Y	
OSW2600	8N	6AC7, 6AC7W	
OSW2601	8Y	6AG7, 6AG7Y	
0SW3104	8R	6SA7, 6SA7G, 6SA7GT, 6SA7GTX, 6SA7GTY, 6SA7Y	
OSW3105	8Q	6SQ7, 6SQ7G, 6SQ7GT	
OSW3106	7AC	6V6, 6V6G, 6V6GT, 6V6GTX, 6V6GTA, 6V6GTY, 6V6GX, 6V6Y	
OSW3107	5L	5AR4, 5AR4/GZ34	5CG4, 5Z4
OSW3109	7Q	6H6, 6H6G, 6H6GT, 6H6MG	

Original	Base Diagram	Preferred Substitution	Good Substitution
OSW3110	6R	6E5	
OSW3111	8N	6SK7, 6SK7G, 6SK7GT, 6SK7GTX, 6SK7GTY 6SK7Y, 6SK7W, 6SK7WA, 6SK7WGT	
OSW3112	6Q	6J5, 6J5G, 6J5GT, 6J5GTX, 6J5GX, 6J5MG	
P12/250			PX4, PP3/250, AC044
P27/250			PX25, PP4/500, D044
P174	5AW	807	
PA5021	4P	866A	
PABC80			9AK8
PC86			4CM4, 4T2
PC88			4DL4
PC93			4BS4
PC95	7FP	4GK5	4FY5
PC97	7FP	4GK5	4FY5
PC900	7GM	4HA5, 4HM5	
PCC18	9A	7AU7	
PCC84			7AN7, 7CC40, 30L1
PCC85	9DE	9AQ8, 9AQ8/PCC85	
PCC88			7DJ8
PCC89			7FC7
PCC186	9A	7AU7	
PCC189			7ES8, 7T29
PCC805			7EK7, B349, 30L15
PCC806			30L17
PCE80			30FL13

Original	Base Diagram	Preferred Substitution	Good Substitution
PCE82			30FL12
PCE800			30FL1, 9GB8, LN339
PCF80	9DC	9A8, 9A8/PCF80	8A8, 30C1, LZ329
PCF82	9AE	6U8, 6U8A, 6KD8	9Q8
PCF86	9MP	7HG8, 7HG8/PCF86	
PCF87			30C17
PCF200	10K	8X9	
PCF201	10K	8U9	
PCF800			30C15, 9EN7, LZ339
PCF801	9QA	8GJ7, 8GJ7/PCF801	
PCF802	9AE	9JW8, 9JW8/PCF802	
PCF805	9KN	7GV7	30C18
PCF806	9QA	8GJ7, 8GJ7/PCF801	PCF801
PCF808			30FL4
PCF900	7GM	4HA5, 4HM5	
PCH200			9V9
PCL82	9EX	16A8, 16A8/PCL82	LN369, 30PL12
PCL83			LN309
PCL84	9HX	15DQ8, 15DQ8/PCL84	15TP7
PCL85	9LY	18GV8, 18GV8/PCL85	
PCL86			14GW8
PCL88			LN329, 30PL14
PCL800	9GK	16GK6	
PCL801			13GC8, LN319, 30P1

Original	Base Diagram	Preferred Substitution	Good Substitution
PD500			9ED4
PE81			30F27
PF9	7R	6K7, 6K7G, 6K7GT, 6K7GTX, 6K7MG	
PF86			4CF8
PFL200			7ED7, 30F5, Z329
PH4	8A	6A8, 6A8G, 6A8GT	
PL17	8CS	5644	
PL21	7BN	2D21, 2D21W, 5727	
PL36	8GT	25E5, 25E5/PL36	25F7
PL81			21A6, PL820, N152
PL82			16A5, 16L40, 30P16, N154
PL83			15A6, N153
PL84	9CV	15CW5, 15CW5/PL84	15BQ5, N378, 30P18
PL86			14GW8
PL136			35FV5
PL300			35FN5
PL302			30P19, N389, 25GF5
PL500	9NH	27GB5, 27GB5/PL500	
PL505	9RJ	40KG6, 40KG6A, 40KG6A/PL505	
PL508			17KW6
PL509	9RJ	40KG6, 49KG6A, 40KG6A/PL505	PL505
PL521	9RJ	29KQ6, 29KQ6/PL521	

Original	Base Diagram	Preferred Substitution	Good Substitution
PL800			16KG8
PL801			12FB5
PL802			16LD8
PL820			21A6, N152, PL81
PL1267	4V	0A4G	
PLL80			12HU8
PM04	7BK	6BA6, 6BA6/EF93, 5749	
PM05	7BD	6AK6, 6AK5/EF95, 5654	
PM07	7BΓ	6AM6, 6AM6/EF91	
PM84			9FG6
PM95	7BK	6AK6	
PP3/250			PX4, AC044, P12/250
PP4/500			D024, PX25, P27/250
PP60	7AC	6L6, 6L6A, 6L6G, 6L6GA, 6L6GB, 6L6GC, 6L6GT, 6L6GX, 6L6GX, 6L6Y, 6L6W, 6L6WA, 6L6WB, 6L6WGT	N66, 5932
PX4			P12/250, PPS/250, AC044, PP3/250
PX25			D024, PP4/250, P27/250
PY31			25Y4, U31
PY32			U291
PY80			19X3, 19W3
PY81	9CB	17Z3, 17Z3/PY81	17R7, U153

Original	Base Diagram	Preferred Substitution	Good Substitution
PY82			19Y3, 19Y40, U154, U192
PY83	9CB	17Z3, 17Z23/PY81	PY81, 20Y40, U329, U251
PY88	9CB	30AE3, 30AE3/PY88	
PY301			19CS4, U339, U191
PY500	6EC4	42EC4, 42EC4/PY500	
PY800	9CB	17Z3, 17Z3/PY81	
PY801	9CB	17Z3, 17Z3/PY81	U193, U349
PZ30			R14
Q160-1		4-125A	
Q400-1		8438, 4-400A	
QA2400	7BK	6065	
QA2401	6BG	6135, 6C4, 6C4W, 6C4WA	
QA2404	6BT	5726, 6AL5	
QA2406	9A	12AT7, 12AT7/ECC81, 12AT7WA, 12AT7WB	
QA2407	9A	6201	
QA2408	8BD	5692	
QB2		250	813
QB3-5		750	6156
QB3-200		8165, 4-65A	
QB3-300		6155	
QB3-300A		4-125A	
QB3.5-750		6156	
QB3.5750GH		5D22, 4-250A	
QB4-250B		5D22, 4-250A	

Original	Base Diagram	Preferred Substitution	Good Substitution
QB4-1100GA		8438, 4-400A	
QB5/1750			6079
QB65	8BD	6SN7, 6SN7A, 6SN7GTA, 6SN7GTB, 6SN7GTY, 6SN7L	
QB309	9A	12AT7, 12AT7/ECC81, 12AT7WA, 12AT7WB	
QBL400		4X500A, 800	
QE03/10	9K	5763	
QE05/40	7CK	6146	
QE05/40H	7CK	6159	
QE06/50	5AW	807	
QE61/250		7203, 4XC250B	
QEL1/150		7034, 4X150A	
QEL1/150H		7609	
QEL2/200		7580	
QEL2/275		7203, 4CX250B	
QF408	1AD4	1AD4	
QL77	6BG	6C4, 6C4W, 6C4WA	
QM328	9G	5686	
QM556	5BS	6X4, 6X4W	
QM557	7BD	5654, 5654W	
QM558	7CM	5725, 6AS6	
QM559	6BT	5726, 6AL5	
QQC04/14			5895
QQE02/5	9HL	6939	
QQE03/12	9PW	6360	
QQE03/12	9PW	6360	
QQE03/20			6252
QQE06/40			5894
QQV02-6	9HL	6939	

Original	Base Diagram	Preferred Substitution	Good Substitution
QQV03-10	9PW	6340	
QQV03-20A		6252	
QQV07/40			829B
QS75/20			75B1
QS83/3			0G3, 85A2
QS92/10V			7475
QS95/10			150AZ, V987B
QS150/15			150B3
QS150/40	4AJ	0D3, 0D3A	150C3, G150/3D
QS150/45			G180/2M
QS150C1	5BO	0A2, 0A2WA, 6073, 6626	
QS150C2	5BO	0A2, 0A2WA, 6073, 6626	
QS150C3	4AJ	0D3, 0D3A	
QS1200			150B2
QS1205	4AJ	0A3, 0A3A	G75/2D
QS1206	4AJ	0C3, 0C3A	G105/10
QS1207	5BO	0A2, 0A2WA, 6073, 6626	G150/4K, 150C2
QS1208	5BO	0B2, 0B2WA, 6074	
QS1209	5BO	5651, 5651A	
QS1210	5BO	0A2, 0A2WA, 6626, 6073	
QS1211	5BO	0B2, 0B2WA, 6074	
QS2404	6BT	6726, 6AL5	
QS2406	9A	12AT7, 12AT7WA, 12AT7WB, 12AT7/ECC81, 6201, 6679	
QV1-150		7304, 4X150A	
QV1-150D		7609	
QV1-150G		8172, 4X150G	

Original	Base Diagram	Preferred Substitution	Good Substitution
QV2-250G		7203. 4CX250B	
QV03-12	9K	5763	
QV05-25	5AW	807	4Y25. 5B/250A
QV06-20	7CK	6146	
QV06-20B	7CK	6883	
QV06-20C	7CK	6159	
QW77			6CQ6
QY2-100		813	
QY2/250		813	
QY3-65A		8165. 4-65	
QY3-125		6155	
QY3-125B		4-125A	
QY4-250		6156	
QY4-250B		5D22. 4-250A	
QY4-400		7527. 4-400B	
QT4-500A		4X500A	
QZ77	7BD	6AM6. 6AM6/EF91	
R3	5BZ	1W4	
R10		6305	HR1
R12		6X2	U43. EY51
R14			PZ30
R16		1T2	U37
R17		6157	
R18		6374	
R19	9Y	1X2. 1X2A. 1X2B. 1X2C.	1R6. DY80
R20		2J2	U49. U26. KY80
R52	5T	5AZ4. 5Y3. 5Y3G. 5Y3GA. 5Y3GT	U50
R144	7BD	6AM6. 6AM6/EF91	
RFY25			ORP12
RL21	7BN	2D21. 2D21W	
RL1267	4V	0A4G	

Original	Base Diagram	Preferred Substitution	Good Substitution
RS2	5L	5Z4	5Y3, 5Y3G, 5Y3GA, 5Y3GT
RS630		100TH	
RS685		4-125A	
RS1002A		5D22, 4-250A	
RS1007		4-125A	
RS1026		5867A	
RS1029	7BD	6AM6, 6AM6/EF91	
RS2016		8170, 4CX5000A	
RS2793		8170, 4XC5000A	
RS4791		8168, 4CX1000A	
RY12-100		8020, 100R	
S6F12	7BD	6AM6, 6AM6/EF91	
S856	5BO	0A2, 0A2WA, 6073, 6626	
S860	5BO	0B2, 0B2WA, 6074	
S901C	5BO	5651	
SM150-30	5BO	0A2, 0A2WA, 6073, 6626	
SP6	7BD	6AM6, 6AM6/EF91	
SR2	5BO		0G3
SR3	4AJ	0B3	
SR55	5BO	0B2, 0B2WA, 6074	
SR56	5BO	0A2, 0A2WA, 6073, 6626	
STR85/10	5BO		0G3
STR108/30	5BO	0B2, 0B2WA, 6074	
STR150/30	5BO	0A2, 0A2WA, 6626, 6073	
STV85/10	5BO	5651, 5651A	0G3

Original	Base Diagram	Preferred Substitution	Good Substitution
STV108/30	5BO	0B2. 0B2WA. 6074	
STV150/30	5BO	0A2. 0A2WA. 6626. 6073	
SU61			6X2
T2M05	7BF	6J6. 6J6A. 6J6WA. 6J6WB. 7244. 6101	
T6D			2B35
T77	6F	6C6	
T130-1		100TH	
T150-1		150TH	
T300-1		450TH	
T380-1		8163. 3-400Z	
T866A	4P	866A	
T1000-1		8164. 3-1000Z	
TAW12-35		6696A	
TB2.5/300			5866
TB3/350		100TH	
TB4/800		250TH	
TB750		5867A	
TD1-100A		7289	
TD03-10			5861. DET22. EC55
TH4327		4EZ7A. 5-125B	
TH5021B	4P	866	
TM12	7BQ	6J4. 6J4WA. 8532	
TS229	9H	5687	
TT10			813
TT16		4-125A	
TT16D		6155	
TT21	6AM	7623	
TT22	6AM	7624	
TTZ63	7R	6J7. 6J7G. 6J7GT. 6J7GTX. 6J7MG	

Original	Base Diagram	Preferred Substitution	Good Substitution
TX2/3	4BZ	5544	
U25			2L2. KY50
U26			2J2
U31			25Y4. PY31
U37	6AR	1T4	1T2. R16
U41	3C	1G3. 1B3. 1B3GT. 1G3GT. 1G3GTA	
U43			6X2. R12. EY51
U47			2L2. KY50
U49			2J2. KY80
U50	5T	5AZ4	R52
U51	5T	5W4. 5W4GT	
U52	5DA	5T4. 5AW4. 5CG4. 6ARA. 5AR4/G234	
U54	5DA	5AR4. 5AR4/GZ34	
U70	6S	6X5	EZ35
U74	5AA	35Z4GT	
U77	5DA	5AR4. 5AR4/GZ34	
U78	5BS	6X4. 6X4W. 6202	6Z31. EZ90
U82	5AB		7Y4
U118			31A3. UY41
U119			38A3. UY85
U142			UY42
U143			AZ31
U145			UY42
U147	6S	6X5. 6X5GT	EZ35
U149	5AB		7Z4
U150			6BT4. EZ40. UU9
U151			6W2. R12. EX81
U152			19U3

Original	Base Diagram	Preferred Substitution	Good Substitution
U153	9CB	17Z3, 17Z3/PY81	
U154			19Y3, PY82
U191			19CS4, PY301
U192			19Y3, 19Y40, PY82
U193	9CB	17Z3, 17Z3/PY81	PY801
U201			CY31
U250	9CB	17Z3, 17Z3/PY81	
U251	9CB	17Z3, 17Z3/PY81	PY83
U291			PY32
U309			19X3, PY80
U319			19R3
U329			PY83
U339			19CS4, PY301
U249	9CB	17Z3, 17Z3/PY81	PY801
U381			38A3, UY85
U404			31A3, UY41
U707	5BS	6X4, 6X4W, 6202	
U709	9DJ	6BW4, 6CA4	EZ91
U718			6BT4, EZ40, UU9
UAA91	6BT	12AL5	10D2
UABC80			28AK8, DH109, 10LD12
UAF41			D2121
UAF42	8BL		12S7, WD142
UB91	6BT	12AL5	10D2
UBC41			14L7, DH118, 10LD3
UBC80			14G6
UBC81			15BD7A, 14G6, DH119, 10LD13

Original	Base Diagram	Preferred Substitution	Good Substitution
UBF80		17C8	
UBF89		19DC8	WD119, 10FD12
UC88			12DL4
UC92			9AB4
UC95			10ER5
UCC84			21CW7
UCC85			28AQ8, B109, 10L14
UCC88			21DJ8
UCC89			22FC7
UCC189			21ES8
UCF80			27BL8
UCH21			UCH71
UCH41			CF141
UCH42			14K7, X142
UCH43			14K7, X142
UCH71			UCH21
UCH80			14Y7
UCH81			19AJ8, X119, 10C14
UCL82	9EX	50GM8, 50GM8/UCL82	48A8, LN119, 10PL12
UCL84			45DQ8
UF41			12AC5, W118, 10F9
UF42			Z142, 10F3
UF80			19BW7
UF85			19BY7
UF89			12DA6
UF183			19EH7
UF184			19EJ7
UL41			45A5, BF451, N142
UL84	9CV		45B5, N119, 10P18

Original	Base Diagram	Preferred Substitution	Good Substitution
UM34			12CD7
UM35			19G5, 10M2
UM80			19BR5, Y119
UM81			19DA5
UM84			19GF6
UN954	5BB	954	
UN955	5BC	955	
UQ80			12BE7
UU9			6BT4, EZ40, U150
UU12	9DJ	6BW4, 6CA4	U709, EZ81
UX866	4P	866A	
UY41			31A3, U118, U404
UY42			31A3, U142, U404
UY82			55N3
UY85			38A3, U119, U381
UY89			31AV3
UY807	5AW	807	
V2M70	5BS	6X4, 6X4W, 6202	
V51			GZ40
V61			6BT4
V153	9CB	17Z3, 17Z3/PY81	
V177			6CQ6
V312			U142, UY42
V311			U118, U404, UY41
V741	6BG	6C4	
V884			6CQ6
V866			6AM5
V987B			QS9510, 150A2
VH550H	4P	866A	

Original	Base Diagram	Preferred Substitution	Good Substitution
VFT6	6R	6U5	
VP6			6CQ6
VP12D	8E		12C8
VR75	4AJ	0A3, 0A3A	
VR75/30	4AJ	0A3, 0A3A	
VR90	4AJ	0B3	
VR105	4AJ	0C3, 0C3A	6105/1D, QS1206
VR105/30	4AJ	0C3, 0C3A	6105/1D, QS1206
VR150	4AJ	0D3, 0D3A	G150/3D, QS150/40
VT83	4C	83	
VT138	6R	1629	6E5
VT139	4AJ	0D3, 0D3A	KD25, CV216, QS150/40
VT202	7BS	9002	
VT203	7BS	9003	
W17	6AR	1T4	1F3, DF91
W25	6AR		1AJ4, 1F1, DF96
W61	7R	5732, 6K7, 6K7G, 6K7GT, 6K7GTX, 6K7MG	
W63	7R	5732, 6K7, 6K7G, 6K7GT, 6K7GTX, 6K7MG	
W76	7R	12K7, 12K7GT	
W77			6CQ6, 9D6, 6F31
W81	8V	7A7	
W118	7CV	12AC5	10F9, UF41
W119			12BX6, 13EC7, 10F18
W143	8V	7H7	
W145	7CV	12AC5	10F9
W147	7R	6K7, 6K7G,	9D4, EF39

Original	Base Diagram	Preferred Substitution	Good Substitution
		6K7GT, 6K7GTX, 6K7MG	
W148	8V	7H7	
W149	8V	7AG7	
W719	9AQ	6BY7	EF8055, 6F19
W727	7BK	6BA6, 6BA6/EF94, 5749	
W739			6EC7, 6F18
WD119			19DC8, 10DF12, UBF89
WD142	8CB	12S7	UAF42
WD150			6CT7, EAF42
WD709	9T	6N8	EBF80
WT210-0001	7BN	2D21, 2D21W, 5727	EN91, E1955
WT210-0003	6Q	884	
WT210-0004	6BS	2050, 2050A	
WT210-0006	7Q	6H6, 6H6G, 6H6GT, 6H6MG	
WT210-0007	7AC	6L6, 6L6A, 6L6G, 6L6GA, 6L6GB, 6L6GC, 6L6GX, 6L6Y, 6L6W, 6L6WA, 6L6WB, 6L6WGT	
WT210-0011	4AJ	0C3, 0C3A	CU686
WT210-0018	4AJ	0D3, 0D3A	KD25, QS150/40, CV216
WT210-0019	4C	83	WT301
WT210-0021	6S	6X5, 6X5G, 6X5GT, 6X5L, 6X5MG, 6X5W, 6X5WGT	
WT210-0028	7AQ	3Q5, 3Q5G, 3Q5GT	

Original	Base Diagram	Preferred Substitution	Good Substitution
WT210-0029	6Q	6C5, 6C5G, 6C5GT	
WT210-0040	5BS	6X4, 6X4W	
WT210-0042	5T	5Y3G, 5Y3GA, 5Y3GT	
WT210-0048	5T	5U4, 5U4G, 5U4GB	
WT210-0060	4R	0Z4, 0Z4A	0Z4G
WT210-0077	7BN	2D21, 2D21W, 5727	
WT210-0081	8N	6SJ7, 6SJ7GT, 6SJ7GTX, 6SJ7GTY, 6SJ7Y, 5693	
WT210-0082	7AC	6V6, 6V6G, 6V6GT, 6V6GTA, 6V6GTX, 6V6GTY, 6V6GX, 6V6Y	
WT210-0084	8B	6N7, 6N7G, 6N7GT, 6N7MG	
WT210-0085	7BZ	50B5	
WT210-0087	8K	6K8, 6K8G, 6K8GT, 6K8GTX	
WT210-0088	6Q	6J5, 6J5G, 6J5GT, 6J5GTX, 6J5GX, 6J5MG	
WT210-0090	6F	6C6, 77	
WT210-0091	4V	0A4G	
WT210-0108	8BD	6AS7, 6AS7G, 6AS7GA, 6080, 6080WA	
WT210-0108	6S	6X5, 6X5G, 6AS7GA, 6080, 6080WA	
WT210-0148	6S	6X5, 6X5G, 6X5GT, 6X5L, 6X5MG, 6X5W, 6X5WGT	

Original	Base Diagram	Preferred Substitution	Good Substitution
WT210-3000	7BN	2D21, 2D21W, 5727	
WT245	6Q2	884	
WT246	6BS	2050, 2050A	
WT261	7Q	6H6, 6H6G, 6H6GT, 6H6MG	
WT261A	7Q	6H6, 6H6G, 6H6GT, 6H6MG	
WT269	4AJ	0C3, 0C3A	
WT294	4AJ	0D3, 0D3A	
WT301	4C	83	
WT301A	4C	83	
WT308	6S	6X5, 6X5G, 6X5GT, 6X5L, 6X5MG, 6X5W, 6X5WGT	
WT389	7AQ	3Q5, 3Q5G, 3Q5GT,	
WT390	6Q	6C5, 6C5G, 6C5GT	
WT606	7BN	2S21, 2D21W, 5727	
WTT100	5BS	6X4, 6X4W	
WTT102	5T	5Y3G, 5Y3GA, 5Y3GT	
WTT106	7Q	6H6, 6H6G, 6H6GT, 6H6MG	
WTT108C1	5BO	0B2, 0B2WA, 6074	
WTT114	4R	0Z4, 0Z4A	0Z4G
WTT122	8N	6SJ7, 6SJ7GT, 6SJ7GTX, 6SJ7GTY, 6SJ7Y, 5693	
WTT123	7AC	6V6, 6V6G, 6V6GT, 6V6GTA, 6V6GTX, 6V6GTY, 6V6GX, 6V6Y	

Original	Base Diagram	Preferred Substitution	Good Substitution
WTT124	7BT	6AT6	
WTT125	8B	6N7, 6N7G, 6N7GT, 6N7MG	
WTT126	7BZ	50B5	
WTT128	8K	6K8, 6K8G, 6K8GT, 6K8GTX	
WTT129	6Q	6J5, 6J5G, 6J5GT, 6J5GTX, 6J5GX, 6J5MG	
WTT131	6F	6C6, 77	
WTT132	4V	0A4G	Z300T, RL1267
WTT135	5T	5U4, 5U4G, 5U4GB	
X14	7Z	1A7	DK32
X17	7AT	1R5	DK91, 1C1
X18			1AC6
X20			1AC6, DK92, 1C2
X25			1AB6, DK96, 1C3
X61M	8K	6K8, 6K8G, 6K8GT, 6K8GTX	
X63	8A	6A8, 6A8G, 6A8GT	
X64	8B	6L7, 6L7G	
X65	8K	6E8	6C31, ECG33
X71M	8K	12K8	
X73M	8A	6D8, 6D8G	6J8
X76M	8K	12K8	
X77	7CH	6BE6, 5750	EK90
X79			6AE8
X81M	8BL	7S7	
X107	7CH	18FX6, 18FX6A	
X118			10C1
X119			19AJ8, 10C14, UCH81
X142			14K7, UCH43

Original	Base Diagram	Preferred Substitution	Good Substitution
X143			ECH21, ECH71
X145			10C1
X147			ECH33, ECH71, 6E8
X148	8BL	7S7	
X155	9AJ	6BZ8, 6BC8	
X719	6E	6AJ8, 6AJ8/ECH81	6C12, EC481, 6CH40
X727	7CH	5750, 6BE6	EK90
XAA91	6BT	3AL5	
XB91	6BT	3AL5	
XC88			2DL4
XC95	7FP	2ER5, 2GK5, 2FQ5, 2FQ5A	2FY5
XC97	7FP	2FY5, 2GK5, 2FQ5, 2FQ5A	2ER5
XC900	7GM	2HA5, 2HM5	
XCC82	9A	7AU7	
XCC89			4FC7
XCC189	9DE	4ES8, 4ES8/XCC189	
XCF80	9AE	4BL8, 4BL8/XCF80	
XCF82	9AE	5U8	
XCF801	9QA	4GJ7, 4GJ7/XCF801	
XCH81			3AJ8
XCL82	9DX	8BM8	
XCL84			8DX8
XCL85	9LY	9GV8, 9GV8/XCL85	
XCL86			8GW8
XF80			3BX6
XF85			3BY7
XF86	9BJ	2HR8	

Original	Base Diagram	Preferred Substitution	Good Substitution
XF94	7BK	3AU6	
XF183	9AQ	3EH7, 3EH7/XF183	
XF184	9AQ	3EJ7, 3EJ7/XF184	
XL36			13CM5
XL84	9CV	8BQ5	
XL86	9CV	8CW5, 8CW5/XL86	
XL136			17FV5
XL500	9NH	13GB5, 13GB5/XL500	14GB5
XXB	7BW	3C6	
XXD	8AC		14F7
XXFM	8AC	7X7	
XXL	5AC		7A4
XY88	9CB	16AQ3, 16AQ3/XY88	
Y25	8GV		1M3
Y61	6R	6G5, 6U5	6M2, 6T5, EM35
Y63	6R	6G5, 6U5	6M2, EM35
Y64	6R	6G5, 6U5	6M2, 6T5, EM35
Y119			19BR5, UM80
YC88			3DL4
YC95	7FP	3ER5	
YC97	7FP	3FY5, 3GK5	
YCC89			5FC7
YCC189	9AJ	5ES8	
YCF86	9MP	5HG8, 6HG8/LCF86	
YCL84	9HX	10DX8, 10DX8/LCL84	
YCL86	9DX	10GW8	
YF183	9AQ	4EH7, 4EH7/LF183	

Original	Base Diagram	Preferred Substitution	Good Substitution
YF184	9AQ	4EJ7, 4EJ7/LF184	
YL84	9CV	10BQ5	
YL86	9CV	10CW5, 10CW5/LL86	
YL1370	7CK	6146, 8298	
YL1371	7CK	8032, 6883	
YL1372	7CK	6159	
YY88	9CB	22AQ3	
Z14	5Y	1N5, 1N5GT	DF33
Z63	7R	7000, 6J7, 6J7G, 6J7GT, 6J7TGX, 6J7MG	
Z77	7DB	6AM6	5A/160H, 6F11
Z90			EF53, EF50
Z142			10F3, UF42
Z145			10F1
Z150			HF62, 6F13, EF42, 6CU7
Z152	9AQ	6BW7	6F41, 8D6, EF80
Z300T	4V	0A4G	
Z319			6351
Z329			7ED7, 30F5, PF818
Z550M			8453
Z719	9AQ	6BX6	6P6, EF800
Z729	9CQ	6BK8, 5750	8D8, 6F22
Z749	9AQ	6EL7	6F33, EF812
Z900T	4CK	5823	
ZA2	5BB	954	E1F

SECTION 2
TUBE BASE DIAGRAMS

This section contains the base diagrams listed in the second column of *Section 1*. From time to time it may be necessary to check these diagrams for internal connects of a particular tube in question. Although it is not recommended. sometimes it is necessary to use a tube that is not a direct replacement as far as the pin connections go. That is. the tube characteristics are acceptable but the internal pin connections differ. In this case it is necessary to change the appropriate pin connections on the tube socket.

These diagrams may also help you service a piece of electronic gear for which you have no schematic. In this case look in *Section 1* to identify the tube base diagram. then return to this section to locate the proper diagram. You can use the diagram to check for proper plate voltage. cathode ground return. screen grid voltage. and so on. You will probably find many uses for these diagrams in the course of servicing. experimenting. and troubleshooting.

IAD4

IAG5

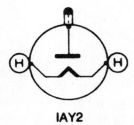

IAY2

IV6

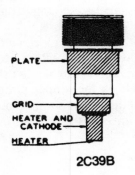

2C39B

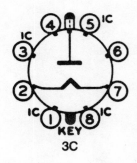

2C40

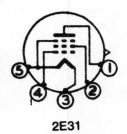

2E31

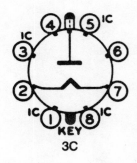

3C

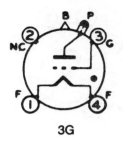

3G

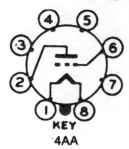

4AA

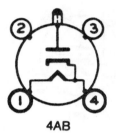

4AB

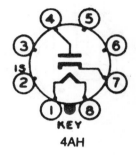

4AH

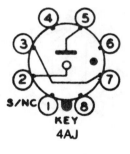

4AJ

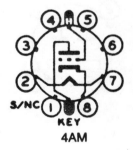

4AM

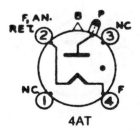

4AT

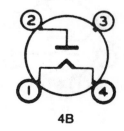

4B

4BJ–4CK

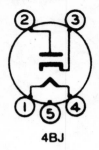

4BJ

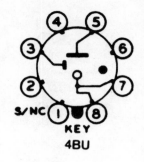

4BU

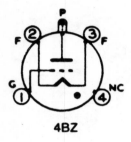

4BZ

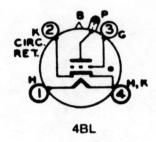

4BL

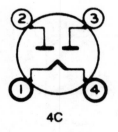

4C

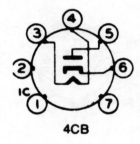

4CB

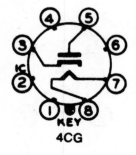

4CG

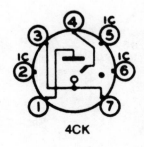

4CK

218

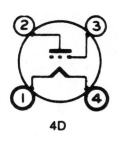

4D

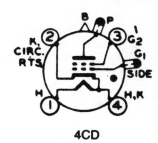

4CD

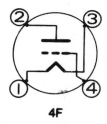

4F

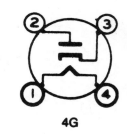

4G

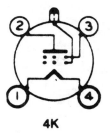

4K

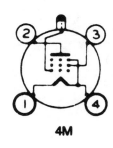

4M

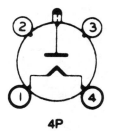

4P

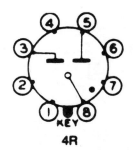

4R

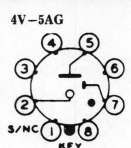

4V

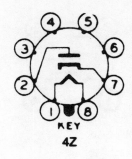

4Z

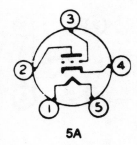

5A

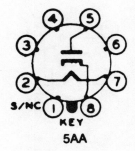

5AA

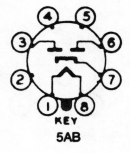

5AB

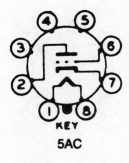

5AC

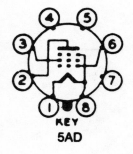

5AD

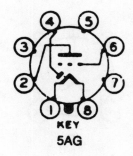

5AG

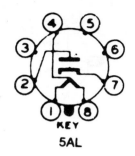

5AL

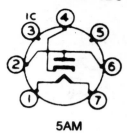

5AM

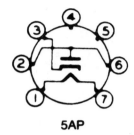

5AP

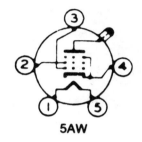

5AW

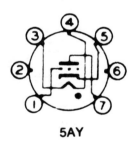

5AY

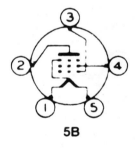

5B

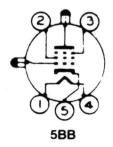

5BB

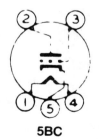

5BC

5BD

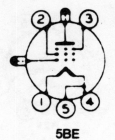

5BE

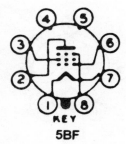

5BF

5BG

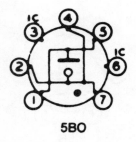

5BO

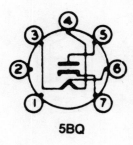

5BQ

5BS

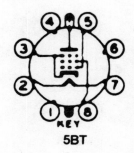

5BT

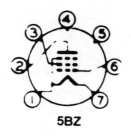

5BZ

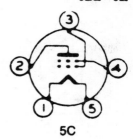

5C

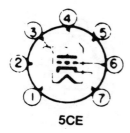

5CE

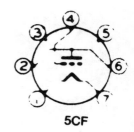

5CF

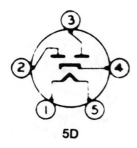

5D

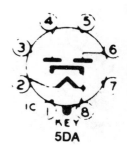

5DA

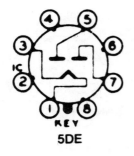

5DE

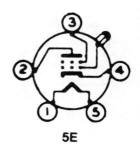

5E

223

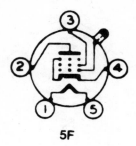

5F

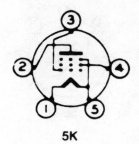

5K

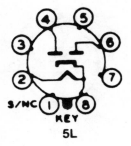

5L

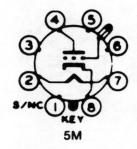

5M

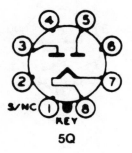

5Q

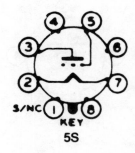

5S

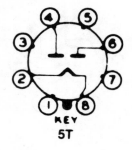

5T

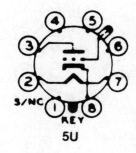

5U

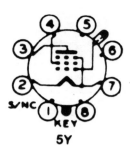

5Y

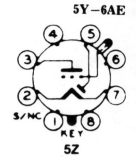

5Z

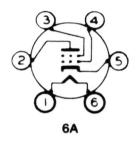

6A

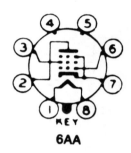

6AA

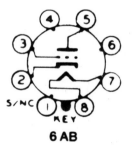

6 AB

6AB8

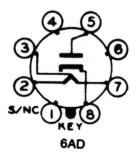

6AD

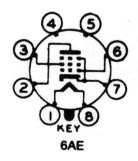

6AE

225

6AF—6BA

6AF

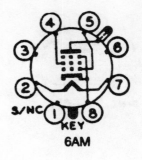

6AM

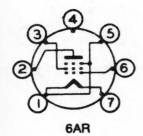

6AR

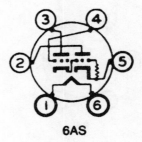

6AS

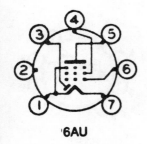

'6AU

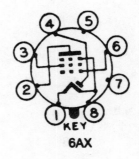

6AX

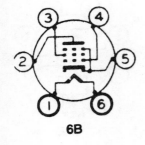

6B

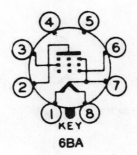

6BA

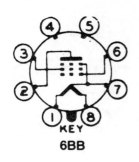

6BB

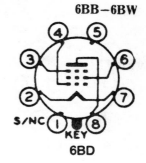

6BB–6BW

6BD

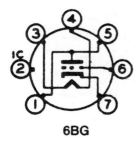

6BG

6BH

6BQ

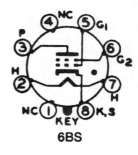

6BS

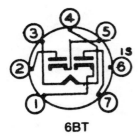

6BT

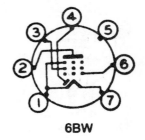

6BW

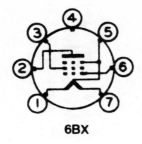

6BX

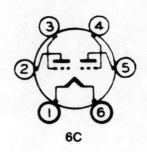

6C

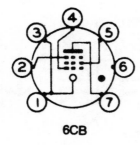

6CB

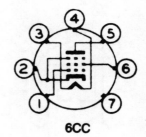

6CC

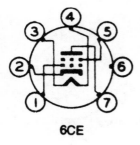

6CE

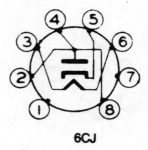

6CJ

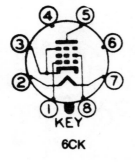

KEY

6CK

6CL

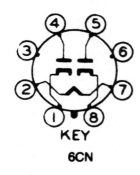

6CN

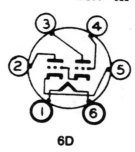

6D

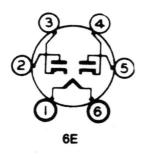

6E

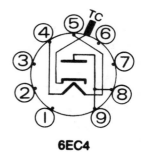

6EC4

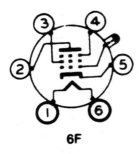

6F

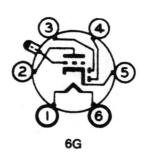

6G

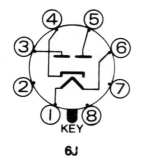

6J

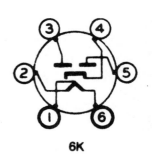

6K

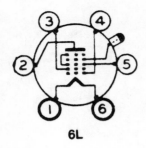

6L

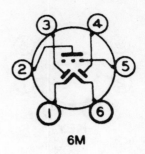

6M

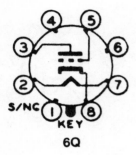

6Q

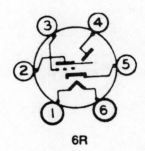

6R

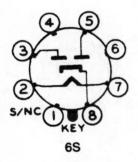

6S

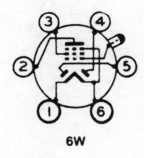

6W

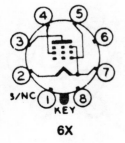

6X

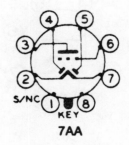

7AA

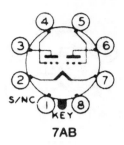

7AB

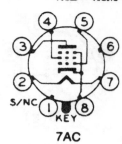

7AC

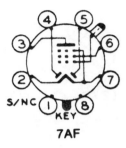

7AF

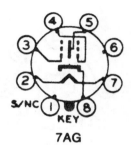

7AG

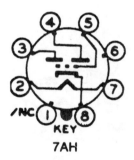

7AH

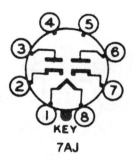

7AJ

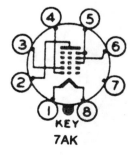

7AK

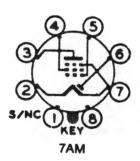

7AM

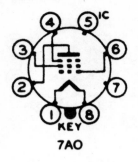

7AO

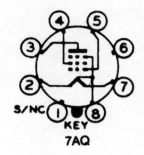

7AQ

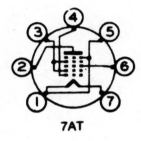

7AT

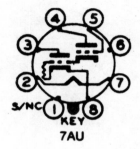

7AU

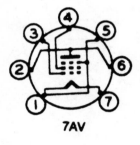

7AV

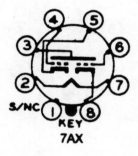

7AX

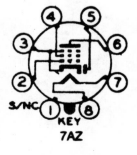

7AZ

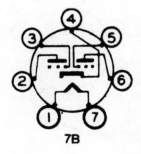

7B

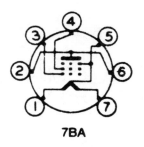

7BA

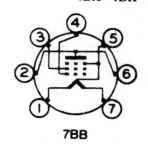

7BB

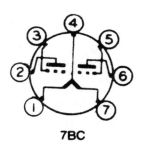

7BC

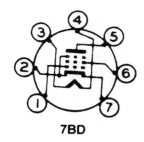

7BD

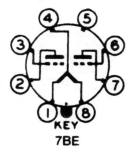

7BE

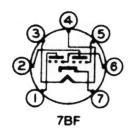

7BF

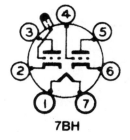

7BH

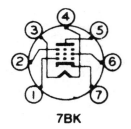

7BK

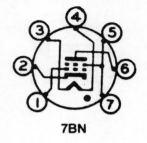

7BN

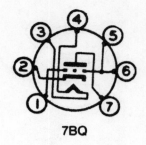

7BQ

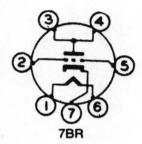

7BR

7BS

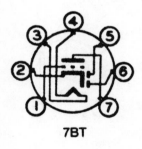

7BT

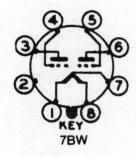

KEY
7BW

7BZ

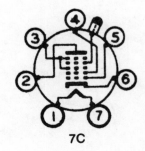

7C

234

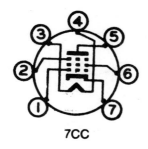

7CC

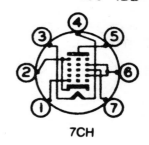

7CH

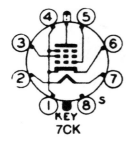

7CK

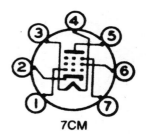

7CM

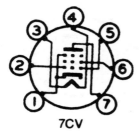

7CV

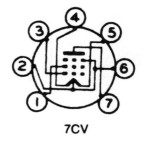

7CV

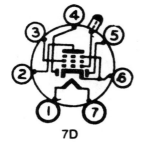

7D

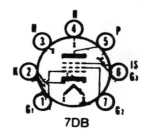

7DB

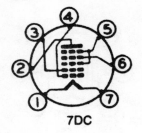

7DC

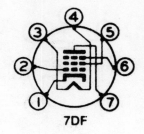

7DF

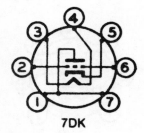

7DK

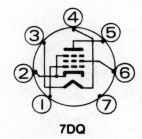

7DQ

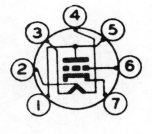

7DW

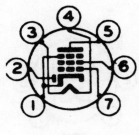

7EA

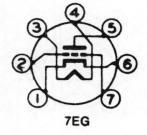

7EG

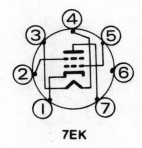

7EK

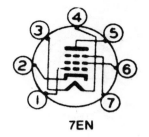

7EN

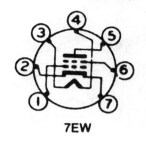

7EW

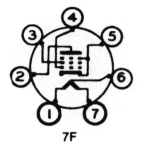

7F

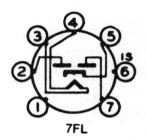

7FL

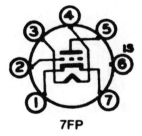

7FP

7FQ

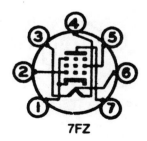

7FZ

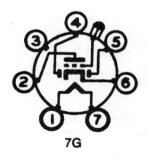

7G

7GA

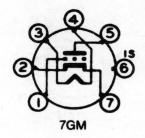

7GM

7H

7K

7Q

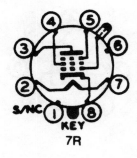

7R

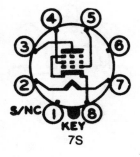

7S

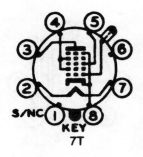

7T

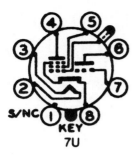

7U

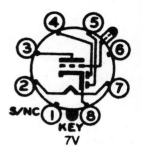

7V

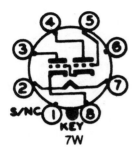

7W

7Z

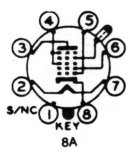

8A

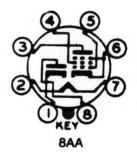

8AA

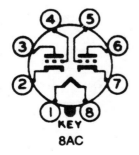

8AC

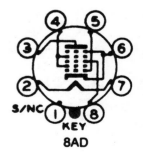

8AD

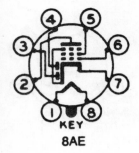

8AE

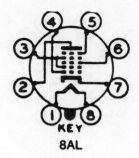

8AL

8AN

8AO

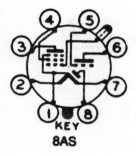

8AS

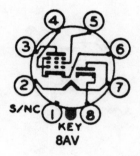

8AV

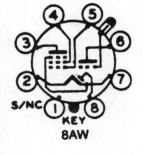

8AW

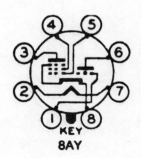

8AY

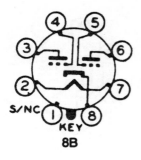

8B

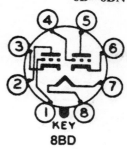

8BD

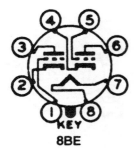

8BE

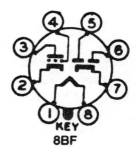

8BF

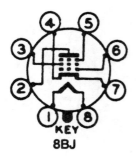

8BJ

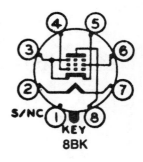

8BK

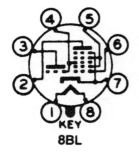

8BL

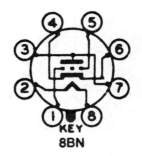

8BN

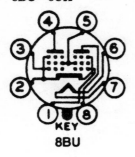

8BU

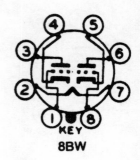

8BW

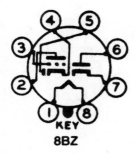

8BZ

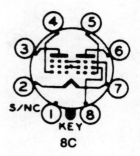

8C

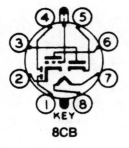

8CB

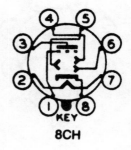

8CH

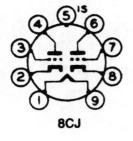

8CJ

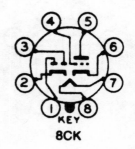

8CK

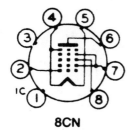

8CN

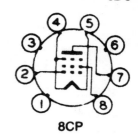

8CP

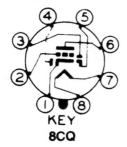

KEY

8CQ

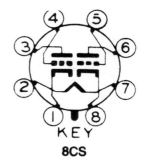

KEY

8CS

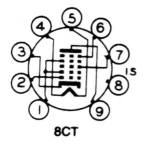

8CT

8DA

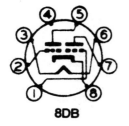

8DB

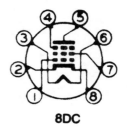

8DC

243

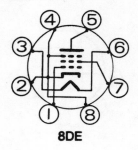

8DE

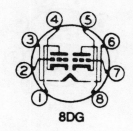

8DG

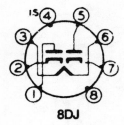

8DJ

8DK

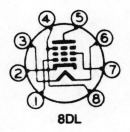

8DL

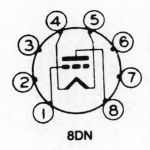

8DN

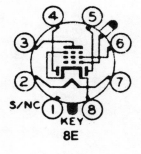

8E

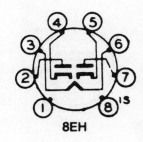

8EH

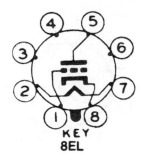

KEY
8EL

KEY
8EP

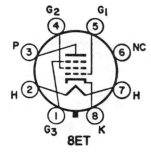

G₂ G₁

P

H

H

G₃ K
8ET

IC IC

IC IC
KEY
8EZ

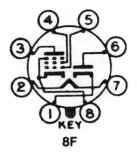

KEY
8F

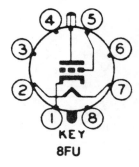

KEY
8FU

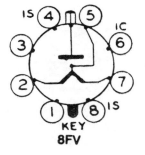

IS IC

IS
KEY
8FV

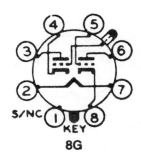

S/NC
KEY
8G

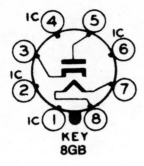

KEY
8GB

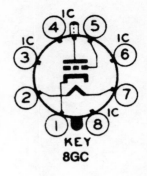

KEY
8GC

KEY
8GD

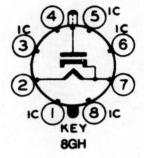

KEY
8GH

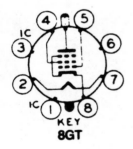

KEY
8GT

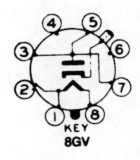

KEY
8GV

KEY
8HY

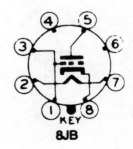

KEY
8JB

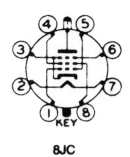

8JC

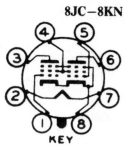

8JP

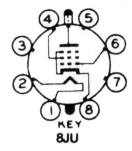

8JU

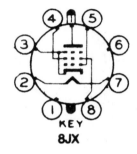

8JX

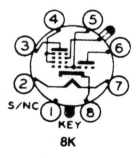

8K

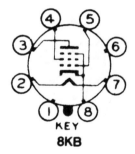

8KB

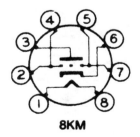

8KM

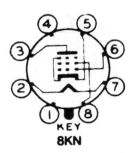

8KN

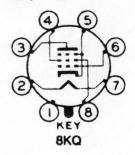

8KQ

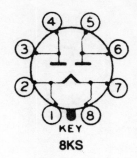

8KS

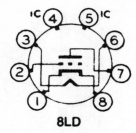

8LD

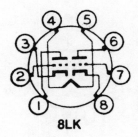

8LK

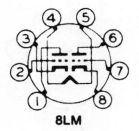

8LM

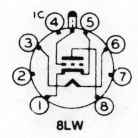

8LW

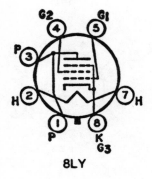

8LY

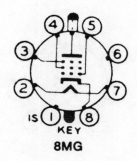

8MG

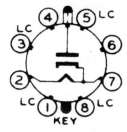

8MH

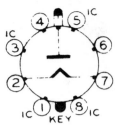

8MK

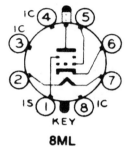

8ML

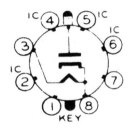

8MT

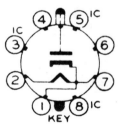

8MU

8MX

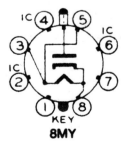

8MY

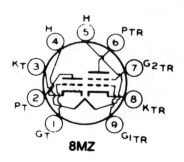

8MZ

249

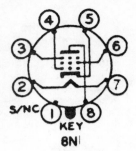

8N

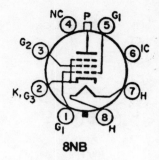

8NB

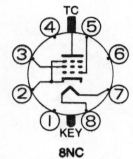

8NC

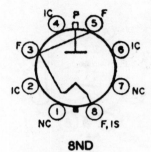

8ND

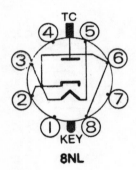

8NL

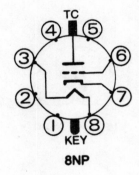

8NP

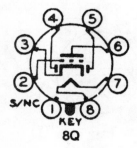

8Q

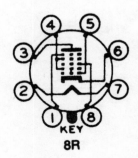

8R

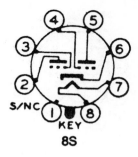

8S

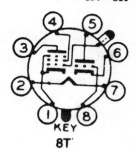

8T

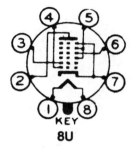

8U

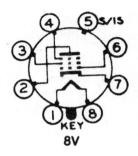

8V

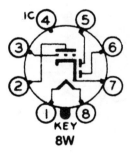

8W

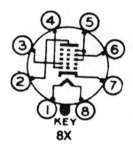

8X

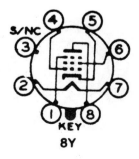

8Y

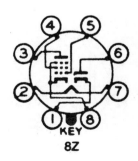

8Z

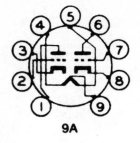

9A

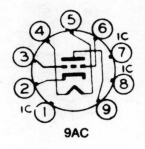

9AC

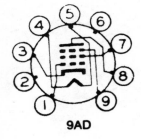

9AD

9AE

9AG

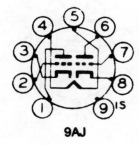

9AJ

9AK

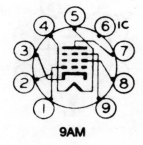

9AM

9AQ–9BQ

9AQ

9AU

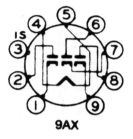

9AX

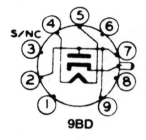

9BD

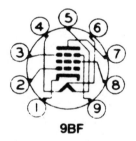

9BF

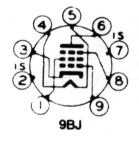

9BJ

9BL

9BQ

253

9BV

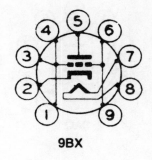

9BX

9CB

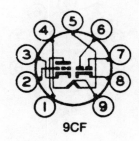

9CF

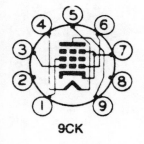

9CK

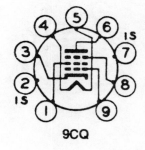

9CQ

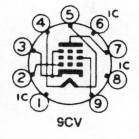

9CV

9CY

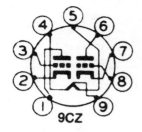

9CZ

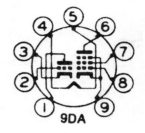

9DA

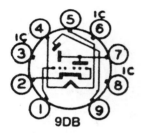

9DB

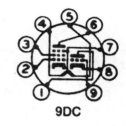

9DC

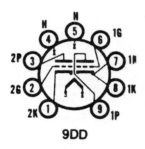

9DD

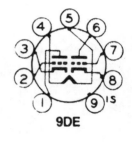

9DE

9DJ

9DP

9DR

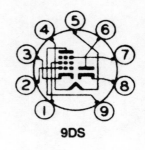

9DS

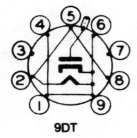

9DT

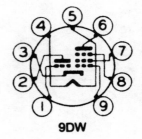

9DW

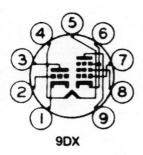

9DX

9DZ

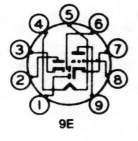

9E

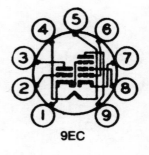

9EC

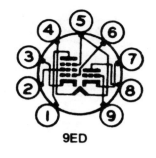

9ED

9EF

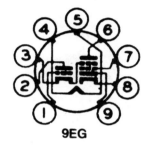

9EG

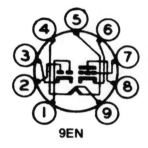

9EN

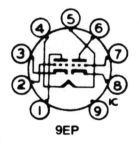

9EP

9EQ

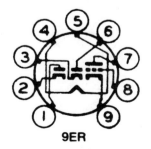

9ER

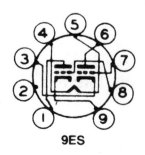

9ES

9EU

9EX

9FA

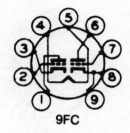

9FC

9FE

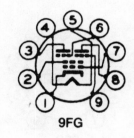

9FG

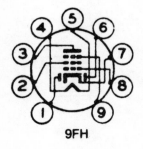

9FH

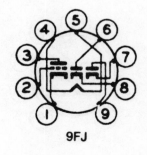

9FJ

9FK

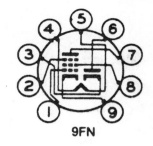

9FN

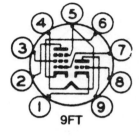

9FT

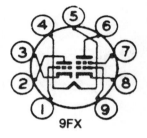

9FX

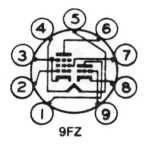

9FZ

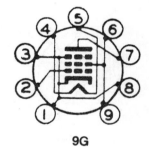

9G

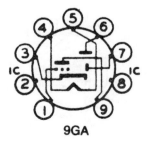

9GA

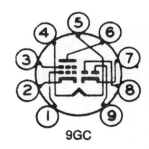

9GC

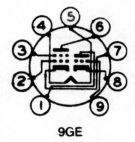

9GE

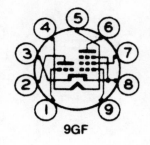

9GF

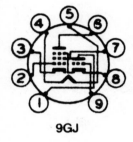

9GJ

9GK

9GM

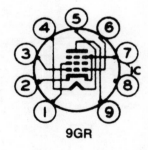

9GR

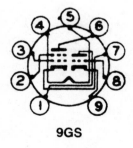

9GS

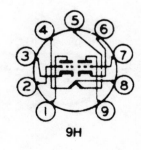

9H

9HE

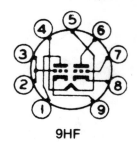

9HF

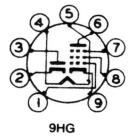

9HG

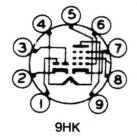

9HK

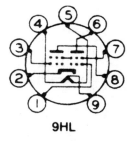

9HL

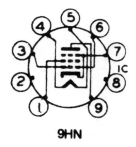

9HN

9HP

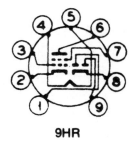

9HR

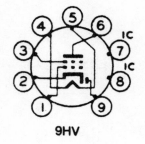

9HV

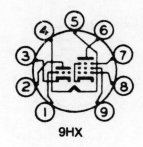

9HX

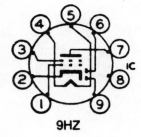

9HZ

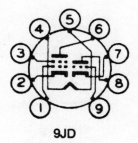

9JD

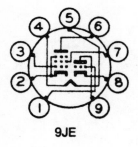

9JE

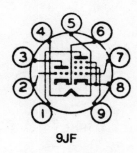

9JF

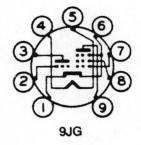

9JG

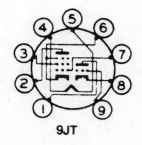

9JT

262

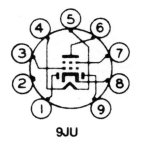

9JU

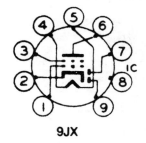

9JX

9K

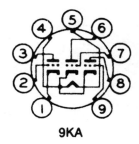

9KA

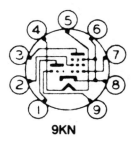

9KN

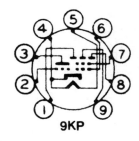

9KP

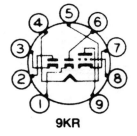

9KR

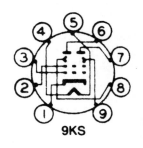

9KS

263

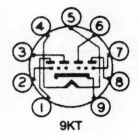

9KT

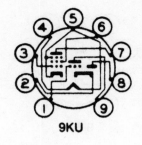

9KU

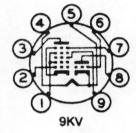

9KV

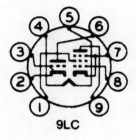

9LC

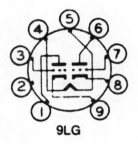

9LG

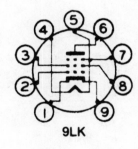

9LK

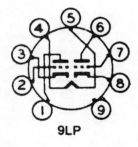

9LP

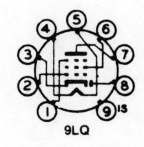

9LQ

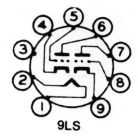

9LS

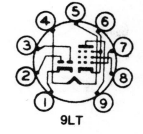

9LT

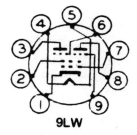

9LW

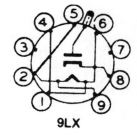

9LX

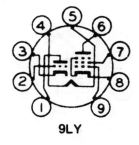

9LY

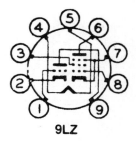

9LZ

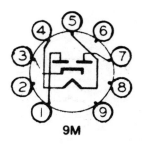

9M

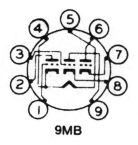

9MB

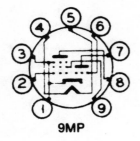

9MP

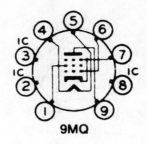

9MQ

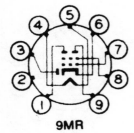

9MR

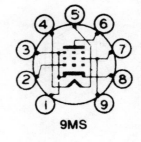

9MS

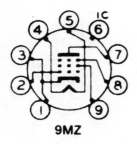

9MZ

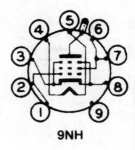

9NH

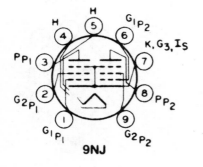

9NJ

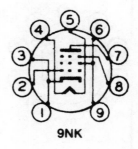

9NK

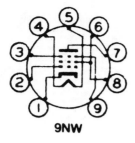

9NW

9NY

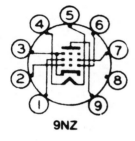

9NZ

9PA

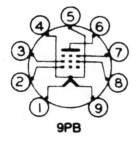

9PB

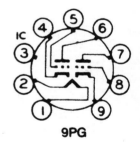

9PG

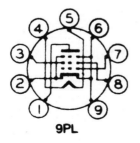

9PL

9PM

9PQ–9QJ

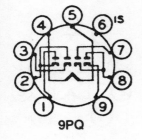

9PQ

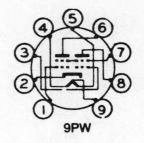

9PW

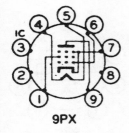

9PX

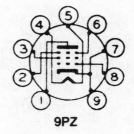

9PZ

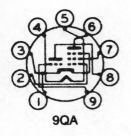

9QA

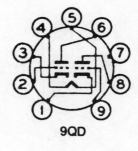

9QD

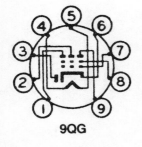

9QG

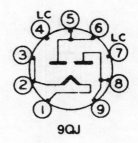

9QJ

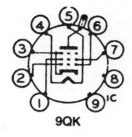

9QK

9QL

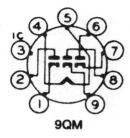

9QM

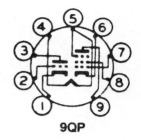

9QP

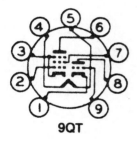

9QT

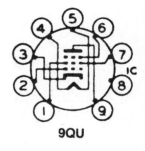

9QU

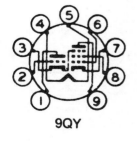

9QY

9QZ

9RA—9RU

9RA

9RF

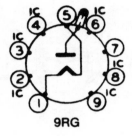

9RG

9RJ

9RL

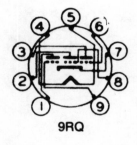

9RQ

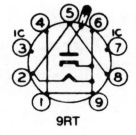

9RT

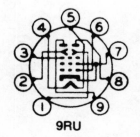

9RU

270

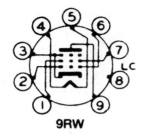

9RW

9RX

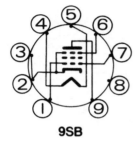

9SB

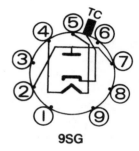

9SG

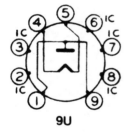

9U

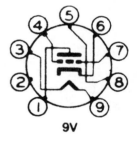

9V

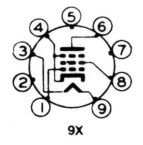

9X

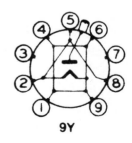

9Y

271

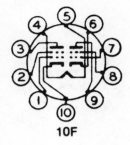

10F

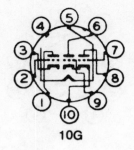

10G

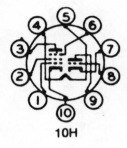

10H

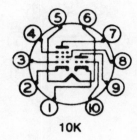

10K

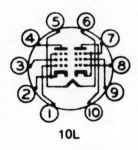

10L

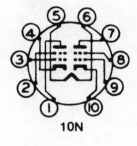

10N

12AQ

12AS

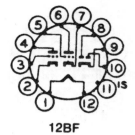

12BF

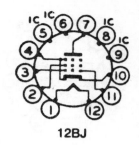

12BJ

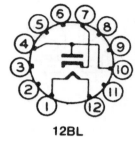

12BL

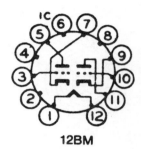

12BM

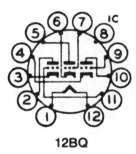

12BQ

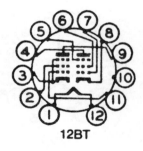

12BT

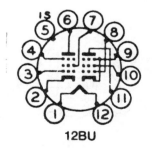

12BU

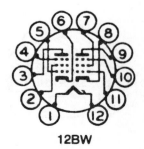

12BW

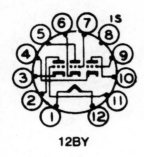

12BY

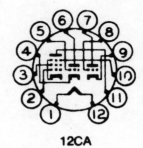

12CA

12CT

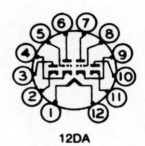

12DA

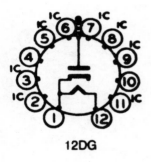

12DG

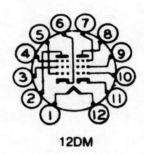

12DM

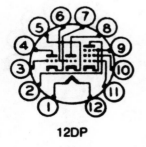

12DP

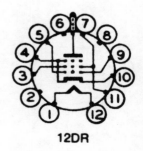

12DR

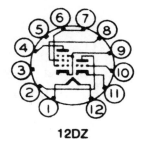

12DZ

12EA

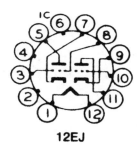

12EJ

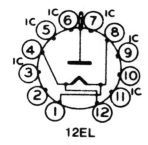

12EL

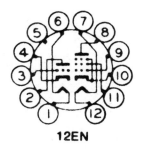

12EN

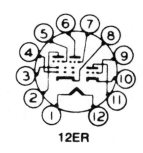

12ER

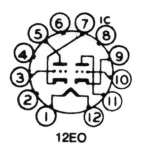

12EO

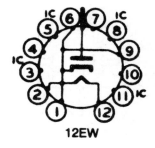

12EW

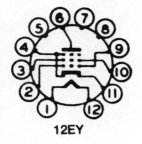

12EY

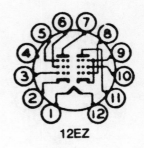

12EZ

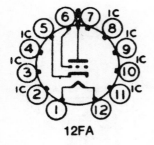

12FA

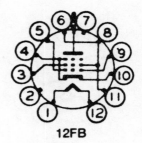

12FB

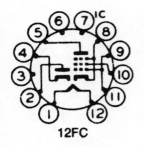

12FC

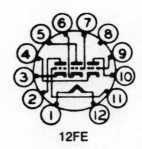

12FE

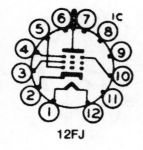

12FJ

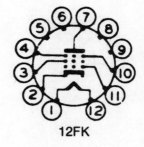

12FK

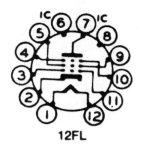

12FL

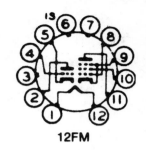

12FM

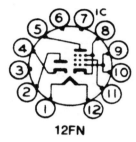

12FN

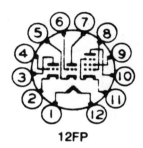

12FP

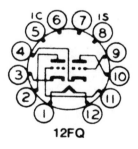

12FQ

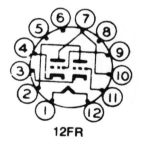

12FR

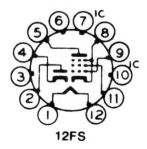

12FS

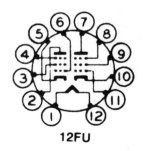

12FU

12FV−12GH

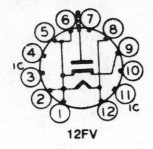

12FV

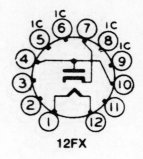

12FX

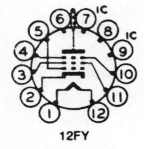

12FY

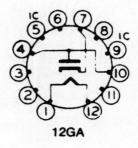

12GA

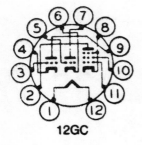

12GC

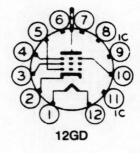

12GD

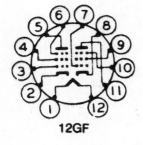

12GF

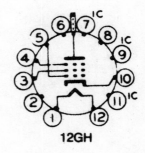

12GH

12GK–12GZ

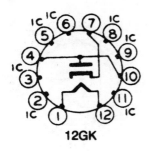

12GK

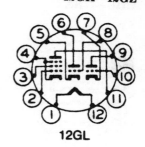

12GL

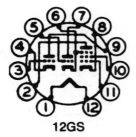

12GS

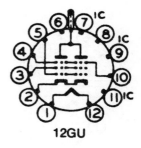

12GU

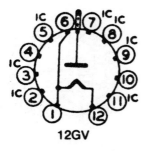

12GV

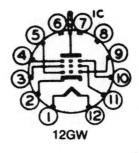

12GW

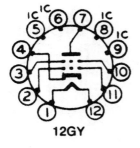

12GY

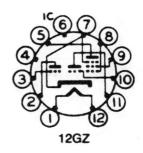

12GZ

279

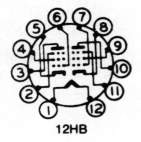

12HB

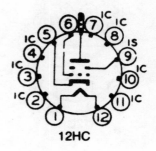

12HC

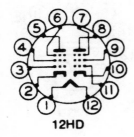

12HD

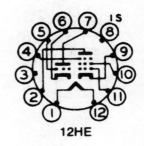

12HE

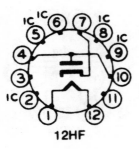

12HF

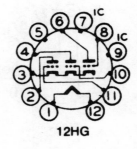

12HG

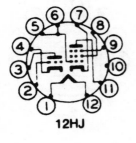

12HJ

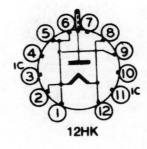

12HK

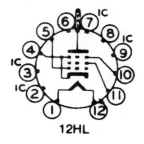

12HL

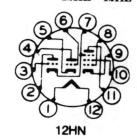

12HN

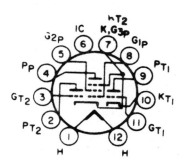

12HR

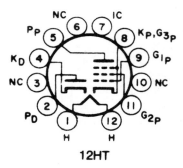

12HT

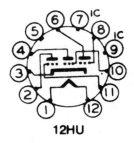

12HU

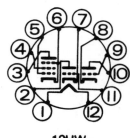

12HW

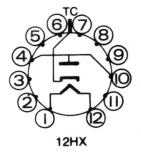

12HX

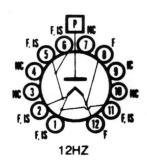

12HZ

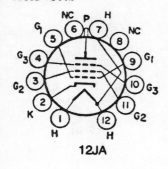

12JA

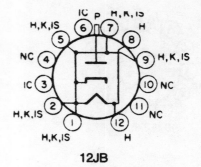

12JB

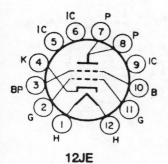

12JE

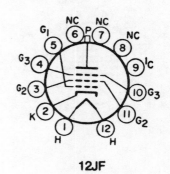

12JF

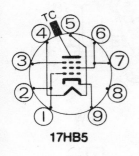

17HB5

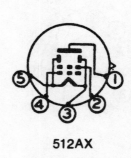

512AX

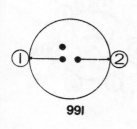

991

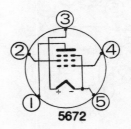

5672

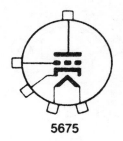

5675

5676

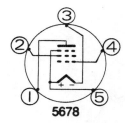

5678

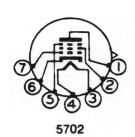

5702

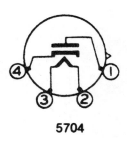

5704

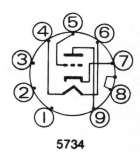

5734

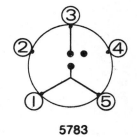

5783

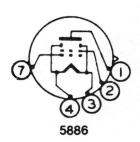

5886

6977 – 8950

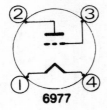

6977

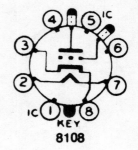

8108

8808

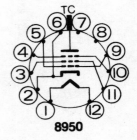

8950

284

SECTION 3
MONOCHROME AND COLOR PICTURE TUBE SUBSTITUTIONS

This section covers picture tube substitutions for both monochrome and color. Monochrome or black-and-white replacements are listed first with the color tubes following. The first column contains the original CRT (cathode-ray tube). the second column lists the replacement or substitute. and the third column indicates the replacement notes necessary for installation. Be sure to read the notes listed under the third column because in some cases it is necessary to make modifications to the receiver chassis or cabinet.

In some cases you will find that some types will cross to another tube but won't cross back. This is because the replacement is a better tube than the original or the neck length may be too long. Nevertheless, you may find that some replacements may cross back that were not indicated. This is due to the *universal style* written into this material. In other words, these substitutions are given for *all* cases—not specific models of receivers. There should be no problem with installation if the appropriate notes are adhered to.

NOTES
1. Direct replacement
2. Mounting hardware may have to be modified
3. Remove safty glass from receiver
4. Matrix replacement
5. Red cathode lead may have to be interchanged with green or blue to obtain gray scale

6. Replacement neck length is slightly longer
7. Receiver mechanical modifications may be required
8. Replacement may not be feasible in small-cabinet recievers—some modifications to mounting hardware may be necessary
9. Use Motorola conversion kit (1U68674A10) when using this replacement
10. Does not require ion trap
11. CRT ground coating must be connected to chassis ground
12. Replacement has 6.3V. 600 mA heater—the receiver CRT heater circuit must be modified
13. Pin 6 (focus anode) must be connected to pin 11 at the socket
14. Replacement is electrically interchangeable—receiver CRT socket must be changed to an RCA 112579 socket or equivalent
15. Replacement is electrically interchangeable—receiver socket must be rewired as follows: connect lead on pin 3 to pin 2 (or pin 6). and connect the lead on pin 6 to pin 3

MONOCHROME SUBSTITUTIONS

Original	Replacement	Notes
5VABP4		
8DP4		
9ACP4	9WP4	1
9AEP4	9VABP4	1
9VABP4		
9VAJP4		
9VAQP4	9VAJP4	1
9WP4		
10ASP4	9AEP4. 9VABP4	1
10ATP4		
10AVP4		
11CP4		
11GP4		
12BFP4	12VAQP4. 12VAWP4	2, 3
12CFP4	12VAQP4. 12VAWP4	1
12CNP4	12VAQP4. 12VAWP4	1
12CNP4A	12VAQP4. 12VAWP4	1
12CTP4	12DKP4	1
12CTP4A	12DKP4	1
12CYP4	12DKP4	1
12DEP4	12DEP4	6

Original	Replacement	Notes
12DFP4		
12DKP4		
12DSP4		
12VAFP4	12VAWP4	1
12VAGP4		
12VAMP4		
12VAQP4	12VAWP4	1
12VAUP4	12VAGP4	1
12VAWP4		
12VAXP4		
12VBNP4		
12VBUP4	12VBNP4	1
12VBYP4	12VBNP4	1
15VACP4		
16BKP4	15VACP4	6
16BTP4	15VACP4	7
16CHP4	15VACP4	1
16CHP4A	15VACP4	1
16CVP4	15VACP4	6, 7
16KP4	16RP4B	10
16KP4A	16RPAB	10
16QP4	16RP4B	10, 11
16RP4	16RP4B	10
16RP4A	16RP4B	10
16TP4	16RP4B	1
16UP4	16RP4B	6, 10, 11
16VAGP4		
16VALP4	16VAGP4	1
16VAQP4	16VAGP4	1
16VBAP4	16VAGP4	1
16VBXP4	16VBYP4	1
16VBYP4		
16XP4	16RP4B	10, 11
17AP4	17BP4D	6, 10, 11
17BP4	17BP4D	10, 11
17BP4A	17BP4D	10
17BP4B	17BP4D	10
17BP4C	17BP4D	10

Original	Replacement	Notes
17BP4D		
17CFP4		
17CYP4	17CFP4	1
17DAP4	17DRP4	1
17DQP4		
17DRP4		
17EMP4	16VAGP4	1
17EWP4	16VAGP4	1
17FDP4	16VAGP4	1
17JP4	17BP4D	10
17LP4	17LP4B	10
17LP4A	17LP4B	10
17LP4B		
17QP4	17QP4B	10
17QP4A	17QP4B	10
17QP4B		
17SP4	17LP4B	10. 13
17UP4	17QP4B	10
17VP4	17LP4B	10
17VP4B	17LP4B	10
17YP4	17QP4B	10
18VAUP4		
18VBLP4	18VAUP4	6
19ABP4		
19ACP4	19CHP4	1
19ADP4	19AVP4	1
19AFP4		
19AGP4	19AVP4	6
19AHP4	19AYP4	1
19AJP4		
19AKP4	19AVP4	1
19ANP4	19AYP4	6
19ARP4	19AFP4	1
19ATP4	19AFP4	6
19AUP4	19AFP4	1
19AVP4		
19AWP4	19AYP4	6
19AXP4	19AYP4	1

Original	Replacement	Notes
19AYP4		
19AZP4	19AVP4	1
19BDP4		
19BHP4	19AVP4	1
19BLP4	19AVP4	6
19BMP4	19AFP4	6
19BRP4	19DRP4	7
19BSP4	19AVP4	6
19BTP4	19AVP4	6
19BVP4	19AVP4	1
19BWP4	19AYP4	1
19BXP4	19AYP4	7
19CAP4	19AVP4	6
19CKP4	19CXP4	1
19CFP4	19CHP4	6, 7
19CHP4		
19CJP4	19AVP4	1
19CKP4	19CHP4	7
19CLP4	19BDP4	1
19CMP4		
19CMP4A	19CMP4	1
19CQP4	19CXP4	1
19CRP4	19BDP4	1
19CSP4	19CHP4	1
19CUP4	19CMP4	1
19CVP4		
19CXP4		
19CYP4	19AVP4	6
19CZP4	19DQP4	7
19DAP4	19DQP4	7
19DBP4		
19DCP4	19DRP4	1
19DEP4	19AVP4	7
19DFP4	19CHP4	1
19DHP4	19DSP4	1
19DKP4	19DRP4	7
19DLP4	19DHP4	1
19DNP4	19DRP4	7

Original	Replacement	Notes
19DQP4		
19DRP4		
19DSP4		
19DTP4	19DQP4	6
19DUP4		
19DWP4		
19DYP4	19CVP4	1
19EAP4	18VAUP4	1
19EBP4		
19EDP4	19DRP4	1
19EFP4	19DSP4	1
19EGP4		
19EHP4	19DRP4	1
19EHP4A	19DRP4	1
19EJP4	18VAUP4	1
19EKP4	19EZP4	1
19ELP4	19AVP4	1
19EMP4	19EBP4	6
19ENP4	18VAUP4	1
19ENP4A	18VAUP4	1
19ERP4	19DRP4	1
19ESP4	19DSP4	1
19EUP4	19DRP4	1
19EVP4	19DQP4	1
19EWP4	19DQP4	1
19EZP4		
19FBP4	19EGP4	1
19FCP4	19DQP4	1
19FCP4A	19DQP4	1
19FDP4	19DQP4	1
19FEP4	18VAUP4	1
19FEP4A	18VAUP4	1
19FEP4B	18VAUP4	1
19FGP4	19EBP4	1
19FJP4	19DQP4	1
19FJP4A	19DQP4	1
19FLP4		
19FNP4	19EBP4	1

Original	Replacement	Notes
19FSP4	18VAUP4	1
19FTP4	19FLP4	1
19FTP4A	19FLP4	1
19FWP4	19AYP4	1
19GAP4		
19GBP4	19DQP4	7
19GEP4	19GEP4A	1
19GEP4A		
19GFP4	19GEP4A	1
19GHP4	19DUP4	6
19GJP4	19DQP4	1
19GJP4A	19DQP4	1
19GMP4	19CVP4	1
19GNP4	19DRP4	1
19GRP4	19DQP4	1
19GTP4	18VAUP4	6
19VAHP4	18VAHP4	1
19VAJP4		
19VALP4		
19VCGP4	19VAHP4	1
19VCJP4	19VAHP4	1
19VCLP4	19VAHP4	1
19VCMP4	19VAHP4	1
19VDEP4	19VALP4	1
19VFEP4		
19XP4	19AVP4	1
19YP4	19AVP4	6
19ZP4	19AVP4	1
20AEP4	19VAHP4	1
20ALP4	19VAJP4	1
20RP4		
20SP4	19VAHP4	1
20UP4	19VALP4	1
20VAQP4		
20YP4	19VAHP4	1
20ZP4	19VAHP4	1
21ACP4	21AMP4B	10
21ACP4A	21AMP4B	10

Original	Replacement	Notes
21AFP4	21YP4B	10. 11
21ALP4	21CBP4A	10. 11
21ALP4A	21CBP4A	10. 11
21ALP4B	21CBP4A	10. 11
21AMP4	21AMP4B	10
21AMP4A	21AMP4B	10
21AMP4B		
21ANP4	21CBP4A	10. 11
21ANP4A	21CBP4A	10. 11
21AQP4	21AMP4B	10. 11
21AQP4A	21AMP4B	10. 11
21ARP4	21ZP4C	1
21ARP4A	21ZP4C	1
21ASP4	21XP4B	10. 11
21ATP4	21CBP4A	10. 11
21ATP4A	21CBP4A	10. 11
21ATP4B	21CBP4A	10. 11
21AUP4	21AVP4C	10
21AUP4A	21AVP4C	10
21AUP4B	21AVP4C	10
21AUP4C	21AVP4C	1
21AVP4	21AVP4C	10
21AVP4A	21AVP4C	10
21AVP4B	21AVP4C	10
21AVP4C		
21AWP4	21AWP4A	10
21AWP4A		
21AYP4	21XP4B	10
21BAP4	21CBP4 A	1
21BCP4	21YP4B	6
21BDP4	21AVP4C	1
21BNP4	21CBP4A	1
21BSP4	21AMP4B	10
21BTP4	21CBP4A	10
21CBP4	21CBP4A	1
21CBP4A		
21CBP4B	21CBP4B	1
21CEP4	21EMP4. 21EQP4	1

Original	Replacement	Notes
21CEP4A	21EMP4, 21EQP4	1
21CMP4	21CBP4A	10
21CQP4		
21CUP4	21AMP4B	10
21CVP4	21CBP4A	1
21CWP4	21CBP4A	10
21CXP4	21DSP4	1
21CZP4	21EMP4, 21EQP4	10
21DAP4	21EMP4, 21EQP4	1
21DEP4	21EMP4, 21EQP4	1
21DEP4A	21EMP4, 21EQP4	1
21DFP4	21EMP4, 21EQP4	1
21DLP4		
21DMP4	21EMP4, 21EQP4	1
21DNP4	21CBP4A	10, 11
21DQP4	21DLP4	1
21DRP4	21CBP4A	1
21DSP4		
21EAP4	21FDP4	12
21EDP4	21EMP4, 21EQP4	1
21EMP4	21EQP4	1
21EP4	21EP4C	10, 11
21EP4A	21EP4C	10
21EP4B	21EP4C	10
21EP4C		
21EQP4	21EMP4	1
21ESP4	21EMP4, 21EQP4	1
21ETP4	21EMP4, 21EQP4	1
21EVP4	21FDP4	6, 12
21FAP4	21EMP4, 21EQP4	1
21FDP4		
21FLP4	21CBP4A	1
21FP4	21FP4D	10, 11
21FP4A	21FP4D	10
21FP4C	21FP4D	10
21FP4D		
21FVP4		
21FWP4	21FVP4	1

Original	Replacement	Notes
21FZP4	21FVP4	1
21GAP4	20VAQP4, 21GAP4A	1
21GAP4A	20VAQP4	1
21JP4	21EP4C	1
21JP4A	21EP4C	1
21KP4	21FP4D	10, 13
21KP4A	21FP4D	10, 13
21MP4	21YP4B	7
21WP4	21WP4B	10
21WP4A	21WP4B	10
21WP4B		
21XP4	21XP4B	10
21XP4A	21XP4B	10
21XP4B		
21YP4	21YP4B	10
21YP4A	21YP4B	10
21YP4B		
21ZP4	21ZP4C	10, 11
21ZP4A	21ZP4C	10
21ZP4B	21ZP4C	10
21ZP4C		
22VABP4		
22CACP4		
22VADP4		
22VAEP4		
22VARP4	22VABP4	1
22VASP4	22VABP4	1
23ACP4	23YP4	1
23AFP4	23YP4	1
23AGP4	23CP4A	6
23AHP4	23ASP4	1
23AKP4	23FP4A	6
23ALP4	23CQP4	1
23ANP4	23BKP4	1
23ARP4		
23ASP4	33AHP4	1
23ATP4	23BKP4	1
23AUP4	23AHP4, 23ASP4	1

Original	Replacement	Notes
23AVP4	23CP4A	6
23AWP4	23BJP4	6
23BAP4	23CP4A	6
23BDP4	23YP4	1
23BFP4	23FP4A	6
23BGP4		
23BHP4	23BGP4	1
23BJP4		
23BKP4		
23BLP4	23BKP4	1
23BMP4	23YP4	1
23BNP4	23CP4A	1
23BP4	23CP4A	6
23BQP4		
23BTP4	23YP4	1
23BVP4	23YP4	1
23BWP4	23YP4	1
23BXP4	23EKP4	7
23BZP4	23CGP4	1
23CBP4	23BQP4	1
23CEP4	23ARP4	1
23CGP4		
23CP4	23CP4A	1
23CP4A		
23CQP4		
23CUP4	23CP4A	6
23CZP4	23AHP4. 23ASP4	1
23DAP4		
23DBP4		
23DKP4	23EKP4	1
23DLP4	23ENP4	6
23DLP4A	23ENP4	6
23DNP4	23BKP4	1
23DP4	23CP4A	6
23DQP4	23BKP4	14
23DSP4	23ENP4	14
23DSP4A	23ENP4	14
23DTP4	23EKP4	1

Original	Replacement	Notes
23DXP4	23CP4A	1
23DYP4	23ETP4	11
23DZP4	23EWP4A	1
23ECP4	23ENP4	7
23EDP4	23EKP4	7
23EHP4	23EKP4	1
23ELP4	23EKP4	1
23EMP4	23EKP4	1
23ENP4		
23EP4		
23EQP4	23EWP4A	1
23ESP4	23HFP4A	1
23ETP4		
23EWP4	23EWP4A	1
23EWP4A		
23EYP4		
23EZP4		
23FBP4	23ENP4	1
23FCP4	22VABP4	1
23FDP4	22VABP4	1
23FEP4	23ENP4	1
23FHP4	22VABP4	1
23FJP4	23ETP4	11
23FLP4	23EKP4	1
23FMP4	23HFP4A	1
23FNP4	22VADP4	1
23FP4	23FP4A	1
23FP4A		
23FRP4		
23FSP4		
23FVP4	23HFP4A	1
23FVP4A	23HFP4A	1
23FVP4B	23HFP4A	1
23FZP4	23FSP4	1
23GBP4	23HFP4A	1
23GEP4	23ENP4	1
23GFP4	22VAEP4	1
23GJP4	22VABP4	1

Original	Replacement	Notes
23GJP4A	22abp4	1
23GP4	23CP4A	1
23GSP4	23FSP4	1
23GTP4	23ETP4	1
23GUP4	23FRP4	1
23Gp4	22VABP4	1
23GWP4		
23GXP4	23FSP4	1
23GZP4	23EKP4	7
23HFP4	23HFP4A	1
23HFP4A		
23HGP4	22VAEP4	1
23HLP4	23FSP4	1
23HP4	23CP4A	1
23HQP4	22VAEP4	1
23HQP4	22VAEP4	1
23HRP4	23HWP4	6
23HSP4	23HWP4A	6
23HUP4	22VABP4	1
23HUP4A	22VAEP4	1
23HWP4	23HWP4A	1
23HWP4A		
23HXP4	23HFP4A	1
23HYP4	23JEP4	1
23JAP4	22VABP4	1
23JBP4	23FSP4	6
23JEP4		
23JEP4	23JEP4	1
23JGP4	23FRP4	1
23JHP4	23HFP4A	1
23JLP4	22VABP4	1
23JP4		
23JRP4	22VACP4	1
23KP4	23FP4A	6
23KP4A	23FP4A	6
23LP4	23ETP4	1
23MP4	23FP4A	1
23MP4A	23FP4A	1

Original	Replacement	Notes
23NP4		
23QP4	23CP4A	1
23TP4	23YP4	1
23UP4	23BQP4	1
23WP4	23FP4A	1
23XP4	23YP4	1
23YP4		
24ADP4	24CP4B	10
24AEP4		
24AHP4		
24ALP4	24AHP4	1
24ANP4	24AEP4	10
24BEP4	24AHP4	15
24CP4	24CP4B	1
24CP4A	24CP4B	10
24CP4B		
24DP4	24AEP4	10
24DP4A	24AEP4	10
24QP4	24CP4B	10, 11
24TP4	24CP4B	10
24VP4	24CP4B	10
24VP4A	24CP4B	10
24XP4	24CP4B	10, 11
24YP4	24AEP4	10
24ZP4	24AEP4	1
230AUB4	9WP4	1
230RB4	9WP4	1
230WB4	9WP4	1
310AVB4	12VAQP4, 12VAWP4	1
310EFB4A	12VAGP4	1
310ENB4	12VAWP4	1
310ERB4	12VAWP4	1
310EVB4	12VAWP4	1
310FCB4	12VAGP4	1
310FJB4	12VAQP4, 12VAWP4	1
440VB4	16VAGP4	1
470ACB4	19AYP4	1
470KB4	19EBP4	1

Original	Replacement	Notes
470NB4	19DRP4	1
470SB4	19DQP4	1
590RB4	23ETP4	1
A31-265W	12VAWP4	1
SG16KP4A	16RP4B	1
SG17BP4B	17BP4B	1
SG17LP4A	17LP4B	1
SG17QP4A	17QP4B	1
SG21ACP4A	21AMP4B	1
SG21AWP4	21AWP4A	1
SG21AWP4A	21AWP4A	1
SG21DEP4A	21EMP4, 21EQP4	1
SG21EP4B	21EP4C	1
SG21FLP4	21CBP4A	1
SG21FP4C	21FP4D	1
SG21WP4A	21WP4B	1
SG21XP4A	21XP4B	1
SG21YP4A	21YP4B	1
SG21ZP4B	21ZP4C	1
SG24AEP4	24AEP4	1
SG24CP4A	24CP4B	

COLOR SUBSTITUTIONS

Original	Replacement	Notes
11SP22	11WP22	1
11WP22		
14VACP22	14VAHP22	1
14VADP22	14VAHP22	1
14VAEP22	14VALP22	1
14VAGP22	14VALP22	1
14VAHP22		
14VALP22		
14VAMP	14VALP22	1
15AEP22	14VALP22	1
15AFP22	14VALP22, 15AEP22	1
15AGP22	14VALP22, 15AEP22	1
15AJP22	14VAHP22	1
15GP22		
15KP22	15LP22	2, 3
15LP22		

Original	Replacement	Notes
15NP22	14VAHP22	1
15RP22	15LP22	2, 3
15SP22	14VALP22, 15AEP22	1
15TP22	14VAHP22, 15NP22	1
15VADTCO1	15VAETCO1	1, 4
15VAETCO1		
15WP22	15LP22	1
15XP22	14VAHP22, 15NP22	1
15ZP22	15LP22	1
16VABP22	16VACP22	1
16VACP22		
16VAHP22	16VACP22	1
17EZP22	16VACP22	1
17FAP22	16VACP22, 17EZP22	1
17FKP22	16VACP22, 17EZP22	1
17VACP22	17VADP22	1, 4
17VADP22	17VACP22	1
17VAMTCO1	17VANTCO1, 17VAYTCO1	1, 4
17VANTCO1	17VAYTCO1	1, 4
17VARP22		
17VAUP22	17VARP22	1, 4
17VAYTCO1	17VANTCO1	1, 4
18VABP22	18VBKP22	1, 4
	18VAHP22	2, 3
18VACP22	18VAHP22	1
18VADP22	18VAHP22	2, 3
	18VBKP22	1, 4
18VAFP22	18VBJP22	1
18VAGP22	18VBDP22	1
	18VBGP22	2
18VAHP22		
18VAJP22	18VBKP22	1, 4
	18VAHP22	2
18VALP22	18VBGP22	1
18VAMP22	18VBDP22	1
	18VBGP22	2
18VANP22		

Original	Replacement	Notes
18VAQP22	18VBKP22	1, 4
	18VAHP22	2
18VARP22	18VAHP22	1
18VASP22	18VBKP22	1, 4
	18VAHP22	2
18VATP22	18VBKP22	1, 4
	18VAHP22	2
18VAZP22		
18VBAP22	18VBKP22	1, 4
	18VAHP22	2
18VBCP22	18VAHP22	1
18VBDP22	18VBGP22	2
18VBGP22		
18VBHP22	18VBKP22	1, 4
	18VAHP22	2
18VBJP22		
18VBKP22		
18VBTP22	18VBKP22	1, 4
	18VAHP22	2
19EXP22	19GVP22	1
	18VAHP22, 19GWP22, 19EYP22	2, 3
19EYP22	18VAHP22, 19GWP22	1
19FMP22	18VAHP22, 19GWP22, 19EYP22	1
19FXP22	18VAHP22, 19GWP22, 19EYP22	1
19GLP22	18VAHP22, 19GWP22, 19EYP22	5
19GSP22	18VAHP22, 19GWP22, 19EYP22	1
19GVP22	19EXP22	1
	18VAHP22	2, 3
19GWP22	18VAHP22, 19EYP22	1
19GXP22	19GVP22, 19EXP22	6
	18VAHP22, 19GWP22, 19EYP22	2, 3, 6
19GYP22	19GVP22, 19EXP22	6
	18VAHP22, 19GWP22, 19EYP22	2, 3, 6
19GZP22	18VAHP22, 19GWP22, 19EYP22	1
19HBP22	18VAHP22, 19GWP22, 19EYP22	1

Original	Replacement	Notes
19HCP22	18VBKP22	1. 4
	19HKP22	1
	18VAHP22. 19GWP22. 19EYP22	2
19HFP22	18VAHP22. 19GWP22. 19EYP22	1
19HJP22	18VBKP22	1. 4
	19HCP22. 19HKP22	1
	18VAHP22. 19GWP22. 19EYP22	2
19HKP22	18VBKP22	1. 4
	19HCP22	1
	18VAHP22. 19GWP22. 19EYP22	2
19HNP22	18VAZP22	1
19HQP22	19HVP22. 19EXP22	1
	18VAHP22. 19GWP22. 19EYP22	2. 3
19HRP22	18VAHP22. 19GWP22. 19EYP22	1
19HTP22	18VBJP22	1
19HXP22	18VBKP22	1. 4
	19HCP22. 19HKP22	1
	18VAHP22. 19GWP22. 19EYP22	2
19HYP22	18VBDP22	1
	18VBGP22	2
19JAP22	18VBDP22	1
	18VBGP22	2
19JBP22	19GVP22. 19EXP22	1
	18VAHP22. 19GWP22. 19EYP22	2. 3
19JCP22	18VBJP22	1
19JDP22	19GVP22. 19EXP22	1
	18VAHP22. 19GWP22. 19EYP22	2. 3
19JGP22	19JWP22	1
19JHP22	18VAHP22. 19GWP22. 19EYP22	1
19JKP22	18VAHP22. 19GWP22. 19EYP22	1
19JNP22	18VBKP22	1. 4
	19HCP22. 19HKP22	1
	18VAHP22. 19GWP22. 19EYP22	2
19JQP22	19GVP22. 19EXP22	1
	19GWP22. 19EYP22. 18VAHP22	2. 3
19JSP22	18VBGP22	1

Original	Replacement	Notes
19JWP22		
19JXP22	18VANP22	1
19JYP22	18VAHP22, 19GWP22, 19EYP22	1
19JZP22	18VBKP22	1, 4
	19HCP22, 19HKP22	1
	18VAHP22, 19GWP22, 19EYP22	2
19KBP22	18VBGP22	1
19KCP22	18VBDP22	1
	18VBGP22	2
19KDP22	18VBJP22	1
19KEP22	18VAHP22	1
19KFP22	18VBKP22	1, 4
	18VAHP22	2
19KHP22	18VBJP22	1
18KJP22	18VBDP22	1
	18VBGP22	2
19VABP22		
19VACP22	19VABP22	1
19VANP22	19VBQP22	1, 4
19VATP22	19VBRP22, 19VCTP22, 19VEDP22	1, 4
	19VDTP22, 19EUP22	1
19VAUP22	19VBRP22, 19VCTP22, 19VEDP22	4, 7
	19VDTP22, 19VEUP22	7
19VBLP22		
19VBQP22		
19VBRP22	19VCTP22, 19VEDP22	1, 4
19VBWP22	19VBRP22, 19VCTP22, 19VEDP22	4, 7
	19VDTP22, 19VEUP22	7
19VCFP22	19VBRP22, 19VCTP22, 19VEDP22	4, 7
19VCNP22	19VBRP22, 19VCTP22, 19VEDP22	1, 4
19VCRP22	19VBRP22, 19VCTP22, 19VEDP22	4, 7
19VCTP22	19VFRP22, 19VEDP22	1, 4
19VCXP22	19VBRP22, 19VCTP22, 19VEDP22	4, 7
19VCYP22	19VBRP22, 19VCTP22, 19VEDP22	4, 7
	19VDTP22, 19VEUP22	7

Original	Replacement	Notes
19VDFP22	19VBRP22, 19VCTP22, 19VEDP22	1, 4
19VDQP22	19VBRP22, 19VCTP22, 19VEDP22	4, 7
19VDSP22		
19VDTP22	19VBRP22, 19VCTP22, 19VEDP22	1, 4
	19VDTP22, 19VEUP22	1
19VDWP22	19VBRP22, 19VCTP22, 19VEDP22	4, 7
	19VDTP22, 19VEUP22	7
19VDXP22	19VBRP22, 19VCTP22, 19VEDP22	4, 7
	19VDTP22, 19VEUP22	7
19VDYP22	19VBRP22, 19VCTP22, 19VEDP22	4, 7
19VEDP22		
19VEKP22	19VEDP22	4, 7
19VELP22	19VEDP22	4, 7
19VESP22	19VEDP22	1, 4
19VEUP22	19VEDP22	1, 4
19VEZP22	19VEDP22	4, 7
	19VEUP22	7
19VFAP22	19VEDP22	4, 7
19VFKP22	19VBRP22, 19VCTP22, 19VEDP22	4, 7
20VABP22	20VAGP22	2
20VADP22	20VAGP22	1
20VAEP22	20VAGP22	2
20VAGP22		
20VAHP22		
20VAJP22	20VAGP22	2
20VAKP22	20VAGP22	2
20VALP22	20VAGP22	2
20VAMP22	20VAGP22	1
20VANP22	20VAGP22	1
20VARP22	20VAGP22	1
20VASP22	20VAGP22	2
21AXP22	21GUP22, 21FBP22A	5, 6, 7
	19VABP22, 21FJP22, 21FJP22A, 21GVP22	2, 3, 5, 6, 7
21AXP22A	21GUP22, 21FBP22, 21FBP22A	5, 6, 7
	19VABP22, 21FJP22, 21FJP22A, 21GVP22	2, 3, 5, 6, 7

Original	Replacement	Notes
21CYP22	21GUP22, 21FBP22, 21FBP22A	5
	21FJP22, 21FJP22A, 21GVP22, 19VABP22	2, 3, 5
21FBP22	21GUP22, 21FBP22A	5
	21FJP22, 21FJP22A, 21GVP22, 19VABP22	2, 3, 5
21FBP22A	21GUP22, 21FBP22	5
	21FJP22, 21FJP22A, 21GVP22, 19VABP22	2, 3, 5
21FJP22	19VABP22, 21FJP22A, 21GVP22	5
21FJP22A	19VABP22, 21FJP22, 21GVP22	5
21FKP22	19VABP22, 21FJP22, 21FJP22A, 21GVP22	5
21GFP22	20VAGP22, 22JP22	8
21GLP22	20VAGP22, 22JP22	8
21GNP22	20VAGP22, 22JP22	8
21GRP22	20VAGP22, 22JP22	8
21GUP22	21FBP22A	1
	19VABP22, 21FJP22, 21GVP22, 21FJP22A	2, 3
21GVP22	19VABP22, 21FJP22, 21FJP22A	1
21GWP22	20VAGP22, 22JP22	8
21GXP22	19VABP22, 21FJP22, 21FJP22A, 21GVP22	1
21GYP22	10VABP22, 21FJP22, 21FJP22A, 21GVP22	1
21GZP22	19VABP22	1
21HAP22	19VABP22	1
21HBP22	20VAGP22, 22JP22	8
21VABP22	21VAKP22	1, 4
21VACP22	21VAZP22, 21VBEP22	1, 4
	21VAKP22	2, 4
21VAFP22	21VAZP22, 21VBEP22	1, 4
	21VAKP22	2, 4
21VAJP22	21VAZP22, 21VBEP22	1, 4
	21VAKP22	2, 4

Original	Replacement	Notes
21VAKP22		
21VAMP22	21VAKP22	1, 4
21VAQP22	21VAZP22, 21VBEP22	1, 4
	21VAKP22	2, 4
21VAUP22	21VAZP22, 21VBEP22	1, 4
	21VAKP22	2, 4
21VAYP22	21VAKP22	1, 4
21VAZP22	21VBEP22	1, 4
	21VAKP22	2, 4
21VBEP22		
21VBFP22	21VBEP22	1, 4
21VBGP22	21VBEP22	1, 4
21VBHP22	21VBEP22	1, 4
22ABP22	20VAGP22, 22JP22	1
22ADP22	22UP22	1
	20VAGP22, 22JP22	2
22AGP22	22UP22	1
	20VAGP22, 22JP22	2
22AHP22	22UP22	1
	20VAGP22, 22JP22	2
22ALP22	22UP22	1
	20VAGP22, 22JP22	2
22AMP22	21VAKP22	1, 4
22ANP22	20VAHP22	1
22ARP22	20VAGP22, 22JP22	1
22ASP22	22UP22	1
	20VAGP22, 22JP22	2
22AVP22	20VAGP22	2
22AXP22	21VAKP22	1, 4
22AYP22	21VAZP22, 21VBEP22	1, 4
	21VAKP22	2, 4
22BCP22	21VAZP22, 21VBEP22	1, 4
	21VAKP22	2, 4
22JP22	20VAGP22	1
22KP22	20VAGP22, 22JP22	2, 3
22LP22	20VAGP22, 22JP22	1

Original	Replacement	Notes
22QP22	20VAGP22, 22JP22	1
22RP22	22KP22	1
	20VAGP22, 22JP22	2, 3
22SP22	20VAGP221, 22JP22	1
22UP22	20VAGP22, 22JP22	2
22XP22	22UP22	1
	20VAGP22, 22JP22	2
22YP22	20VAGP22, 22JP22	1
23EGP22	24VALP22, 25BCP22	4, 9
	23VANP22, 25XP22, 25AP22A	9
23EGP22A	23VALP22, 25BCP22	4, 9
	23VANP22, 25XP22, 25AP22A	9
23VABP22	23VALP22	1, 4
	23VANP22	1
23VACP22	23VAQP22	1
	23VANP22	2
	23VALP22	2, 4
23VADP22	23VAQP22	1
	23VANP22	2
	23VALP22	2, 4
23VAGP22	23VAQP22	1
	23VANP22	2
	23VALP22	2, 4
23VAHP22	23VALP22	1, 4
	23VANP22	1
23VALP22		
23VAMP22	23VALP22	2, 4
23VANP22	23VALP22	1, 4
23VAQP22	23VANP22	2
	23VALP22	2, 4
23VARP22	23VANP22	1
	23VALP22	1, 4
23VASP22	23VALP22	1, 4
23VATP22	23VALP22	1, 4
23VAUP22	23VALP22	1, 4
23VAWP22	23VALP22	2, 4

Original	Replacement	Notes
23VAXP22	23VALP22	1, 4
23VAYP22	23VALP22	2, 4
23VAZP22	23VALP22	2, 3, 4
23VBAP22	23VALP22	2, 4
23VBCP22	23VAQP22	1
	23VANP22	2
	23VALP22	2, 4
23VBDP22	23VAQP22	1
	23VANP22	2
	23VALP22	2, 4
23VBEP22	23VANP22	1
	23VALP22	1, 4
23VBGP22	23VALP22	1, 4
23VBHP22	23VANP22	1
	23VALP22	1, 4
23VBJP22	23VAQP22	1
	23VANP22	2
	23VALP22	2, 4
23VBNP22	23VAQP22	1
	23VANP22	2
	23VALP22	2, 4
23VBRP22	23VAQP22	1
	23VANP22	2
	23VALP22	2, 4
23VBSP22	23VANP22	1
	23VALP22	1, 4
23VBTP22	23VAQP22	1
	23VANP22	2
	23VALP22	2, 4
23VBWP22	23VALP22	2, 4
23VBYP22	23VAQP22	1
	23VANP22	2
	23VALP22	2, 4
23ABP22	23VANP22, 25XP22, 25AP22A	1
	23VALP22	1, 4
25ADP22	23VAQP22	1

Original	Replacement	Notes
	23VANP22, 25XP22, 25AP22A	2
	23VALP22, 25BCP22	2, 4
25AEP22	25YP22, 25BP22A	1
	23VANP22, 25XP22, 25AP22A	2, 3
	23VALP22, 25BCP22	2, 3, 4
25AFP22	23VANP22, 25XP22, 25AP22A	1
	23VALP22, 25BCP22	1, 4
25AGP22	23VAQP22	1
	23VANP22, 25XP22, 25AP22A	2
	23VALP22, 25BCP22	2, 4
25AJP22	23VAQP22	1
	23VANP22, 25XP22, 25AP22A	2
	23VALP22, 25BCP22	2, 4
25ANP22	23VANP22, 25XP22, 25AP22A	1
	23VALP22, 25BCP22	1, 4
25AP22	23VANP22, 25XP22, 25AP22A	1
	23VALP22, 25BCP22	1, 4
25AP22A	23VANP22, 25XP22	1
	23VALP22, 25BCP22	1, 4
25AQP22	23VANP22, 25XP22, 25AP22A	1
	23VALP22, 25BCP22	1, 4
25ASP22	23VAQP22	1
	23VANP22, 25XP22, 25AP22A	2
	23VALP22, 25BCP22	2, 4
25AWP22	23VAQP22	1
	23VANP22, 25XP22, 25AP22A	2
	23VALP22, 25BCP22	2, 4
25AXP22	23VAQP22	1
	23VANP22, 25XP22, 25AP22A	2
	23VALP22, 25BCP22	2, 4
25AZP22	23VAQP22	1
	23VANP22, 25XP22, 25AP22A	2
	23VALP22, 25BCP22	2, 4
25BAP22	23VALP22, 25BCP22	1, 4
25BCP22	23VALP22	1, 4
25BDP22	23VALP22, 25BCP22	2, 4

Original	Replacement	Notes
25BFP22	23VAQP22	1
	23VANP22, 25XP22, 25AP22A	2
	23VALP22, 25BCP22	2, 4
25BGP22	23VANP22, 25XP22, 25AP22A	1
	23VALP22, 25BCP22	1, 4
25BHP22	23VAQP22	1
	23VANP22, 25XP22, 25AP22A	2
	23VALP22, 25BCP22	2, 4
25BJP22	23VANP22, 25XP22, 25AP22A	1
	23VALP22, 25BCP22	1, 4
25BKP22	23VALP22, 25BCP22	2, 4
25BMP22	23VANP22, 25XP22, 25AP22A	1
	23VALP22, 25BCP22	1, 4
25BP22	23YP22, 25BP22A	1
	23VANP22, 25XP22, 25AP22A	2, 3
	23VALP22, 25BCP22	2, 3, 4
25BP22A	25YP22	1
	23VANP22, 25XP22, 25AP22A	2, 3
	23VALP22, 25BCP22	2, 3, 4
25BRP22	23VANP22, 25XP22, 25AP22A	1
	23VALP22, 25BCP22	1, 4
25BSP22	23VAQP22	1
	23VANP22	2
	23VALP22	2, 4
25BVP22	23VANP22	1
	23VALP22	1, 4
25BWP22	23VAQP22	1
	23VANP22	2
	23VALP22	2, 4
25BXP22	23VANP22	1
	23VALP22	1, 4
25BZP22	23VANP22	1
	23VALP22	1, 4
25CBP22	23VANP22	1
	23VALP22	1, 4
25CP22	23VANP22, 25XP22, 25AP22	1
	23VALP22, 25BCP22	1, 4

Original	Replacement	Notes
25CP22A	23VANP22, 25XP22, 25AP22A	1
	23VALP22, 25BCP22	1, 4
25FP22	25YP22, 25BP22A	1
	23VANP22, 25XP22, 25AP22A	2, 3
	23VALP22, 25BCP22	2, 3, 4
25GP22	23VANP22, 5XP22, 25AP22A	1
	23VALP22, 25BCP22	1, 4
25GP22A	23VANP22, 25XP22, 25AP22A	1
	23VALP22, 25BCP22	1, 4
25RP22	25YP22, 25BP22A	1
	23VANP22, 25XP22, 25AP22A	2, 3
	23VALP22, 25BCP22	2, 3, 4
25SP22	23VANP22, 25XP22, 25AP22A	1
	23VALP22, 25BCP22	1, 4
25VABP22	25VCKP22, 25VDAP22, 25VDXP22	1, 4
25VACP22	25VCZP22	1, 4
	25VCKP22, 25VABP22, 25VDAP22, 25VDXP22	2, 4
25VADP22	25VABP22, 25VCKP22, 25VDAP22, 25VDXP22	1, 4
25VAEP22	25VABP22, 25VCKP22, 25VDAP22, 25VDXP22	1, 4
25VAFP22	25VCZP22	1, 4
	25VAEP22	2
	25VCLP22, 25VABP22, 25VDAP22, 25VDXP22	2, 4
25VAGP22	25VBEP22	1, 4
25VAJP22	25VAEP22	1
	25VABP22, 25VCKP22, 25VDAP22, 25VDXP22	1, 4
25VAKP22	25VAEP22	1
	25VABP22, 25VCKP22, 25VDAP22, 25VDXP22,	1, 4
25VALP22	25VABP22, 25VCKP22, 25VDAP22, 25VDXP22	1, 4
25VAMP22	25VABP22, 25VCKP22, 25VDAP22, 25VDXP22	1, 4

Original	Replacement	Notes
25VAQP22	25VCZP22	1, 4
	25VABP22, 25VCKP22, 25VDAP22, 25VDXP22	2, 4
25VARP22	25VAEP22	1
	25VABP22, 25VCKP22, 25VDAP22, 25VDXP22	1, 4
25VASP22	25VAEP22	1
	25VABP22, 25VCKP22, 25VDAP22, 25VDXP22	1, 4
25VATP22	25VAEP22	1
	25VABP22, 25VCKP22, 25VDAP22, 25VDXP22	1, 4
25VAUP22	25VABP22, 25VCKP22, 25VDAP22, 25VDXP22	1, 4
25VAWP22	25VCZP22	1, 4
	25VABP22, 25VCKP22, 25VDAP22, 25VDXP22	2, 4
25VAZP22	25VCZP22	1, 4
	25VAEP22	2
	25VABP22, 25VCKP22, 25VDAP22, 25VDXP22	2, 4
25VBAP22	25VAEP22	1
	25VABP22, 25VCKP22, 25VDAP22, 25VDXP22	1, 4
25VBEP22		
25VBGP22	25VCZP22	1, 4
	25VAEP22	2
	25VABP22, 25VCKP22, 25VDAP22, 25VDXP22	2, 4
25VBHP22	25VAEP22	1
	25VABP22, 25VCKP22, 25VDAP22, 25VDXP22	1, 4
25VBKP22	25VBEP22	1, 4
25VBLP22	25VAEP22	1
	25VABP22, 25VCKP22, 25VDAP22, 25VDXP22	1, 4

Original	Replacement	Notes
25VBMP22	25VABP22. 25VCKP22. 25VDAP22. 25VDXP22	1, 4
25VBNP22	25VCZP22	1, 4
	25VAEP22	2
	25VABP22. 25VCKP22. 25VDAP22. 25VDXP22	2, 4
25VBQP22	25VCZP22	1, 4
	25VABP22. 25VCKP22. 25VDAP22. 25VDXP22	2, 4
25VBRP22	25VCZP22	1, 4
	25VABP22. 25VCKP22. 25VDAP22. 25VDXP22	2, 4
25VBTP22	25VBEP22	1, 4
25VBUP22	25VCZP22	1, 4
	25VABP22. 25VCKP22. 25VDAP22. 25VDXP22	2, 4
25VBWP22	25VCZP22	1, 4
	25VABP22. 25VCKP22. 25VDAP22. 25VDXP22	2, 4
25VBXP22	25VAEP22	1
	25VABP22. 25VCKP22. 25VDAP22. 25VDXP22	1, 4
25VBYP22	25VBEP22	1, 4
25VBZP22	25VCZP22	1, 4
	25VABP22. 25VCKP22. 25VDAP22. 25VDXP22	2, 4
25VCAP22	25VCZP22	1, 4
	25VABP22. 25VCKP22. 25VDAP22. 25VDXP22	2, 4
25VCBP22	25VCZP22	1, 4
	25VABP22. 25VCKP22. 25VDAP22. 25VDXP22	2, 4
25VCDP22	25VBEP22	1, 4
25VCEP22	25VCZP22	1, 4
	25VABP22. 25VCKP22. 25VDAP22. 25VDXP22	2, 4

Original	Replacement	Notes
25VCFP22	25VCZP22	1, 4
	25VABP22, 25VCKP22, 25VDAP22, 25VDXP22	2, 4
25VCGP22	25VABP22, 25VCKP22, 25VDAP22, 25VDXP22	1, 4
25VCKP22	25VDAP22, 25VDXP22	1, 4
25VCNP22	25VCZP22	1, 4
	25VDAP22, 25VDXP22	2, 4
25VCUP22	25VCZP22	1, 4
	25VDAP22, 25VDXP22	2, 4
25VCWP22	25VCZP22	1, 4
	25VABP22, 25VCKP22, 25VDAP22, 25VDXP22	2, 4
25VCZP22	25VDAP22, 25VDXP22	2, 4
25VDAP22	25VDXP22	1, 4
25VDBP22	25VCZP22	1, 4
	25VDAP22, 25VDXP22	2, 4
25VDEP22	25VDAP22, 25VDXP22	1, 4
25VDJP22	25VABP22, 25VCKP22, 25VDAP22, 25VDXP22	1, 4
25VDKP22	25VDAP22, 25VDXP22	1, 4
25VDMP22	25VDZP22	4, 7
	25VDAP22, 25VDXP22	2, 4
25VDNP22	25VCZP22	1, 4
	25VDAP22, 25VDXP22	2, 4
25VDSP22	25VCZP22	1, 4
	25VDAP22, 25VDXP22	2, 4
25VDXP22	25VDAP22, 25VDXP22	1, 4
25VP22	23VANP22, 25XP22, 25AP22A	1
	23VALP22, 25BCP22	1, 4
25WP22	23VANP22, 25XP22, 25AP22A	1
	23VALP22, 25BCP22	1, 4
25XP22	23VANP22, 25AP22A	1
	23VALP22, 25BCP22	1, 4

Original	Replacement	Notes
25YP22	25BP22A	1
	23VANP22, 25XP22, 25AP22A	2, 3
	23VALP22, 25BCP22	2, 3, 4
25ZP22	23VANP22, 25XP22, 25AP22A	1
	23VALP22, 25BCP22	1, 4
26AP22	25VABP22, 25CKP22, 25VDAP22, 25VDXP22	1, 4
26DP22	25VCZP22	1, 4
	25VABP22, 25VCKP22, 25VDAP22, 25VDXP22	2, 4
26FP22	25VAEP22	1
	25VABP22, 25VCKP22, 25VDAP22, 25VDXP22	1, 4
26GP22	25VCZP22	1, 4
	25VAEP22	2
	25VABP22, 25VCKP22, 25VDAP22, 25VDXP22	2, 4
26HP22	25VBEP22	1, 4
26KP22	25VABP22, 25VCKP22, 25VDAP22, 25VDXP22	1, 4
370AB22	14VALP22, 15AEP22	1
370CB22	14VAHP22, 15NP22	1
370EB22	14VALP22, 15AEP22	1
490AB22	19GVP22, 19EXP22	5
	18VAHP22, 19GWP22, 19EYP22	2, 3, 5
490ACB22	19GVP22, 19EXP22	5
	18VAHP22, 19GVP22, 19EYP22	2, 3, 5
490ADB22	19GVP22, 19EXP22	5
	18VAHP22, 19GWP22, 19EYP22	2, 3, 5
490AEB22	18VAHP22, 19GWP22, 19EYP22	5
490AFB22	18VAHP22, 19GWP22, 19EYP22	5
490AGB22	18VAHP22, 19GWP22, 19EYP22	5
490AHB22	19GVP22, 19EXP22	5
	18VAHP22, 19GWP22, 19EYP22	2, 3, 5
490AHB22A	19GVP22, 19EXP22	1
	18VAHP22, 19GWP22, 19EYP22	2, 3

Original	Replacement	Notes
490AHB22B	18VAHP22	2, 3
490AJB22	18VAHP22, 19GWP22, YP22	1
490AJB22A	18VAHP22, 19GWP22, 19EYP22	1
490AJB22B	18VAHP22	1
490AKB22	19GVP22, 19EXP22	5
	18VAHP22, 19GWP22, 19EYP22	2, 3, 5
490AKB22A	19GVP22, 19EXP22	1
490ALB22	19GVP22, 19EXP22	5
	18VAGP22, 19GWP22, 19EYP22	2, 3, 5
490AMB22	19GVP22, 19EXP22	5
	18VAGP22, 19GWP22, 19EYP22	2, 3, 5
490AMB22A	19GVP22, 19EXP22	1
	18VAHP22, 19GWP22, 19EYP22	2, 3
490ANB22	19GVP22, 19EXP22	5
	18VAHP22, 19GWP22, 19EYP22	2, 3, 5
490ARB22	18VAHP22, 19GWP22, 19EYP22	5
490ASB22	18VAHP22, 19GWP22, 19EYP22	1
490ASB22A	18VAGP22, 19GWP22, 19EYP22	1
490ASB22B	18VAHP22	1
490BAB22	19GVP22, 19EXP22	1
	18VAHP22, 19GWP22, 19EYP22	2, 3
490BCB22	18VAHP22, 19GWP22, 19EYP22	1
490BDB22	18VAHP22, 19GWP22, 19EYP22	5
490BDB22C	18VAHP22	1
490BDB22D	18VAHP22	1
490BGB22	19GVP22, 19EXP22	1
	18VAHP22, 19GWP22, 19EYP22	2, 3
490BHB22	18VAHP22, 19GWP22, 19EYP22	1
490BNB22	19JWP22	1
490BRB22	18VAHP22, 19GWP22, 19EYP22	1
490BSB22	19JWP22	2, 3
	18VAZP22	3, 8
490BVB22	19JWP22	1
490BXB22	19JWP22	1
490BZB22	18VAHP22	2, 3
490BZB22A	18VAHP22	2, 3

Original	Replacement	Notes
490CAB22	18VAHP22, 19GWP22, 19EYP22	1
490CAB22A	18VAHP22	1
490CB22	19VP22, 19EXP22	5
	18VAHP22, 19GWP22, 19EYP22	2, 3, 5
490CHB22	19GVP22, 19EXP22	1
	18VAHP22, 19GWP22, 19EYP22	2, 3
490CHP22A	18VAHP22	2, 3
490CSB22A	18VAHP22	2, 3
490CUB22	18VAHP22, 19GWP22, 19EYP22	1
490DB22	18VAHP22, 19GWP22, 19EYP22	1
490DB22A	18VAHP22, 19GWP22, 19EYP22	1
490DEB22	19JWP22	1
490DHB22	18VBDP22	1
	18VBGP22	2
490EB22	19GVP22, 19EXP22	5
	18VAHP22, 19GWP22, 19EYP22	2, 3, 5
490EB22A	19GVP22, 19EXP22	5
	18VAHP22, 19GWP22, 19EYP22	2, 3, 5
490FB22	19GVP22, 19EXP22	5
	18VAHP22, 19GWP22, 19EYP22	2, 3, 5
490GB22	19GVP22, 19EXP22	5
	18VAHP22, 19GWP22, 19EYP22	2, 3, 5
490HB22	19GVP22, 19EXP22	1
	18VAHP22, 19GWP22, 19EYP22	2, 3
490JB22	19GVP22, 19EXP22	1
	18VAHP22, 19GWP22, 19EYP22	2, 3
490JB22A	19VP22, 19EXP22	1
	18VAHP22, 19GWP22, 19EYP22	2, 3
490KB22	19GVP22, 19EXP22	5
	18VAHP22, 19GWP22, 19EYP22	2, 3, 5
490KB22A	19GVP22, 19EXP22	5
	18VAHP22, 19GWP22, 19EYP22	2, 3, 5
490LB22	19GVP22, 19EXP22	5
	18VAHP22, 19GWP22, 19EYP22	2, 3, 5
490MB22	19GVP22, 19EXP22	5
	18VAHP22, 19GWP22, 19EYP22	2, 3, 5

Original	Replacement	Notes
490NB22	18VAHP22, 19GWP22, 19EYP22	5
490RB22	18VAHP22, 19GWP22, 19EYP22	5
490SB22	18VAHP22, 19GWP22, 19EYP22	5
490TB22	18VAHP22, 19GWP22, 19EYP22	5
490UB22	19GVP22, 19EXP22	5
	18VAHP22, 19GWP22, 19EYP22	2, 3, 5
490VB22	18VAHP22, 19GWP22, 19EYP22	5
	18VAHP22, 19GWP22, 19EYP22	2, 3, 5
490XB22	18VAHP22, 19GWP22, 19EYP22	5
490YB22	18VAHP22, 19GWP22, 19EXP22	5
490ZB22	18VAHP22, 19GWP22, 19EYP22	5
510ELB22	19VDTP22, 19VEUP22	1
	19VBRP22, 19VCTP22, 19VEDP22	1, 4
510ERB22	19VBRP22, 19VCTP22, 19VEDP22	4, 7
550AB22A	20VAGP22	2, 3
A49-14X	19GVP22, 19EXP22	5
	18VAHP22, 19GWP22, 19EYP22	2, 3, 5

Notes

Notes

Notes